Qualitative Research in Physical Activity and the Health Professions

William A. Pitney, EdD, ATC

Jenny Parker, EdD

Northern Illinois University

Human Kinetics

D1379176

Library of Congress Cataloging-in-Publication Data

Pitney, William A., 1965-
  Qualitative research in physical activity and the health professions / William A. Pitney and Jenny Parker.
    p. ; cm.
  Includes bibliographical references and index.
  ISBN-13: 978-0-7360-7213-7 (soft cover)
  ISBN-10: 0-7360-7213-6 (soft cover)
  1. Qualitative research. I. Parker, Jenny, 1964- II. Title.
  [DNLM: 1. Exercise. 2. Qualitative Research. 3. Physical Fitness. 4. Research Design. QT 255 P685q
2009]
  R737.P65 2009
  613.7'1072--dc22

                                    2008054464

ISBN-10: 0-7360-7213-6 (print)   ISBN-10: 0-7360-8544-0 (Adobe PDF)
ISBN-13: 978-0-7360-7213-7 (print)   ISBN-13: 978-0-7360-8544-1 (Adobe PDF)

The Web addresses cited in this text were current as of December 11, 2008, unless otherwise noted.

Acquisitions Editor: Loarn D. Robertson, PhD; **Developmental Editor:** Kathleen Bernard; **Assistant Editors:** Jillian Evans and Nicole Gleeson; **Copyeditor:** Joy Wotherspoon; **Proofreader:** Anne Meyer Byler; **Indexer:** Craig Brown; **Permission Manager:** Dalene Reeder; **Graphic Designer:** Nancy Rasmus; **Graphic Artist:** Denise Lowry; **Cover Designer:** Keith Blomberg; **Photographer (interior):** © Human Kinetics, except far right photos on page 1, 29, 83, and 115, which are © Photodisc; **Photo Asset Manager:** Laura Fitch; **Photo Production Manager:** Jason Allen; **Art Manager:** Kelly Hendren; **Illustrator:** Keri Evans; **Printer:** Edwards Brothers Malloy

Printed in the United States of America   10   9   8   7   6   5   4

The paper in this book is certified under a sustainable forestry program.

Human Kinetics
Web site: www.HumanKinetics.com

United States: Human Kinetics, P.O. Box 5076, Champaign, IL 61825-5076
800-747-4457
email: humank@hkusa.com

Canada: Human Kinetics, 475 Devonshire Road Unit 100, Windsor, ON N8Y 2L5
800-465-7301 (in Canada only)
email: info@hkcanada.com

Europe: Human Kinetics, 107 Bradford Road, Stanningley, Leeds LS28 6 AT, United Kingdom
+44 (0) 113 255 5665
email: hk@hkeurope.com

Australia: Human Kinetics, 57A Price Avenue, Lower Mitcham, South Australia 5062
08 8372 0999
e-mail: info@hkaustralia.com

New Zealand: Human Kinetics, P.O. Box 80, Torrens Park, South Australia 5062
0800 222 062
e-mail: info@hknewzealand.com

To my true north: Lisa, Liam, and Quinlan.

–W. Pitney

For Mum and Nancy: You keep me grounded on both sides of the Atlantic and remind me of what is truly important.

–J. Parker

Contents

Preface

This is an exciting time for professionals associated with physical activity and the health professions. We have seen a substantial increase in the amount and type of research being conducted. Qualitative research has grown in popularity and has gained respect as a viable method of answering important research questions. Despite its broad appeal and inclusion in many disciplines, however, qualitative research often has an ethereal or mystical feel to it, particularly for those practitioners with strong roots in physical activity and the health professions.

As qualitative researchers, we have conducted many studies, taught many graduate-level research courses, and advised a great many graduate students in using qualitative methods for their theses and dissertations. Our experiences as educators and researchers give us a unique view of qualitative methods. Our lens is broad in regard to physical activity and health professions, and includes athletic training, physical education, physical therapy, health education, nursing, and general medical disciplines.

You may be asking, "Another research text? Aren't there enough?" Indeed, there is no shortage of research texts. However, an informal appraisal of the textual resources for research classes reveals an interesting paradox. On one hand, though many general research textbooks do a fantastic job of presenting a broad spectrum of research methods, their discussion of the guiding principles and applications of qualitative research is limited. One or two chapters may provide an overview of qualitative methods, but the text lacks a depth of information required to do the topic justice.

On the other hand, numerous texts on qualitative research exist in the disciplines of social science, such as sociology, anthropology, and education. However, most of these texts contain an enormous amount of information, and it is hard for students, practitioners, and novices of qualitative research to sort out and apply the important concepts and procedures. For example, although some texts present introductory content for beginners to qualitative methods, they also include advanced ideology related to qualitative research, such as critical theory and postmodernism. These tangential discussions, though important in their own right, tend to muddy the clarity of the methods for those just learning the ropes of interpretive inquiry. We believe that for many students in physical activity and the health professions, and for practitioners alike, qualitative research methods look very foreign as compared to quantitative research methods that are more common and traditional.

Qualitative Research in Physical Activity and the Health Professions addresses these problems by explaining the underlying principles of qualitative inquiry in a clear manner that helps students and practitioners fully understand how to design, conduct, and evaluate a qualitative research study. We systematically present the content with terms that are consistent with traditional forms of research to reveal how qualitative methods frame a researchable problem, derive purposes and questions from the problem, and guide procedures for data collection and analysis. Additionally, this text includes excerpts from published studies in the chapters and full research articles in the appendices so readers can put principles into practice. For example, when explaining how to create purpose statements and research questions, we provide real samples from scholarly publications. The concise nature, real examples, dialogue boxes, recommended learning activities, and suggested supplemental readings make this a very versatile textbook.

As professionals in the fields of physical activity and the health professions, we function in complex environments and interact with many different people. Because the majority of our work occurs in social contexts, we are constantly prompted to consider the human condition. We are required to make significant decisions and to effectively solve clinical and educational problems. Systematic inquiry certainly guides our professional practice and informs our ability to make decisions. We must be good consumers of research, including the qualitative methods that are now entering the fold. We have written this text to explain the qualitative research principles so practitioners can effectively evaluate published qualitative studies. Our text is organized into four parts.

Part I outlines the characteristics of qualitative research and introduces the general principles that guide this form of inquiry. It also dissects a qualitative study to foreshadow the content of part II.

Part II explains how to conceptualize and conduct a qualitative research study, describing the modes of data collection and analysis, as well as the steps needed to obtain trustworthy data. This part concludes with a discussion of the ethical principles that guide the qualitative research process.

Part III discusses how to write qualitative research. It begins with the process of assembling a research plan and progresses to writing results and discussing findings. The chapter on writing research provides concrete examples of how to present and discuss findings after a study has been completed.

The final section of the text, part IV, contains information for those who plan to continue learning about qualitative research. Chapter 9 discusses the various forms of qualitative methods that researchers and consumers encounter. Chapter 10 focuses on how to evaluate qualitative studies. Chapter 11 contains advice about handling the common challenges and criticisms of qualitative inquiry. This section includes other important aspects of qualitative research, such as how it is combined with traditional methods to form a mixed-methods approach. The chapter concludes with practical advice and resources for pursuing projects in qualitative research.

We have used a variety of pedagogical strategies to facilitate learning. Each part opener introduces its content in both written and graphic form. For example, the following figure illustrates the overall structure of our text.

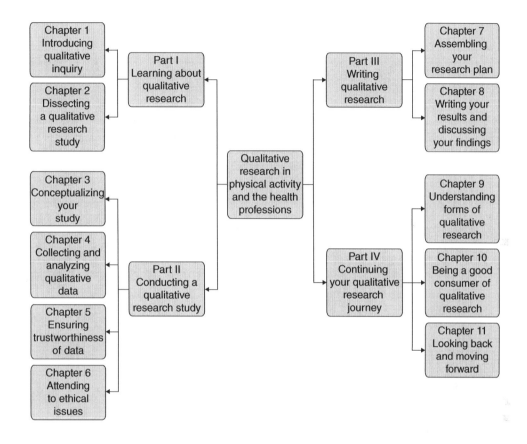

We set out to write a practical text for those who are first learning about qualitative research. We hope you enjoy reading the text as much as we have enjoyed writing it.

Acknowledgments

We would like to thank our students who are the central focus for us as educators. Special thanks to Dr. Loarn Robertson, Kathleen Bernard, and the staff at Human Kinetics for their guidance and support.

–W.A. Pitney and J. Parker

⊙

I would first like to thank my friend, colleague, and writing partner, Jenny Parker. Without her insights and scholarly ability, this text would not have been possible. Jenny, it has truly been my pleasure to work with you; even Dr. Grewant thinks so.

Thank you to my mentor and friend, Paul Ilsley, for starting me on my qualitative research journey.

Special thanks to my colleagues, coauthors, and copresenters with whom I have spent many hours working on various projects: David, Mike, Stacy, Tom, Jen, Marie, Brian, Christine, Shaun, Greg, Craig, Alex, Stephanie, Skip, Lorin, John, Justin, Jim, Moira, Jason, Paul, Jan, Alicia, Paul, Arun, and Tanya. I have learned so much from you. I also appreciate the friendship and support of so many others: Susie, Robyn, Laurie, Beth, Suraj, Pommy, Jan, So-Yeun, Paul, Vicky, Sue, and Nancy. You rock!

To my family, whom I love dearly. You put a smile on my face.

–W.A. Pitney

⊙

Completion of a book involves many people both directly and indirectly. I hope you all know who you are and how much you are appreciated.

In particular I would like to thank Bill Pitney, who shared his vision, time, expertise, and humor throughout the ups and ups (yes, I really mean that!) of the writing and revision process. How fortunate I am to have you as a friend and colleague.

I am very appreciative of Judy, Patt, Larry, Judy, and Linda, who have always gently nudged me to write, write, and rewrite.

I would also like to thank Ethel, So-Yeun, Susie, and Robyn, whose midweek laughs continue to be an essential part of life.

And finally, I thank the people whose encouragement, love, and support know no bounds: Jools, EE, Wee Gracie, Laura, Jackie, Neil, The Serious One, Susan, Margee, Sydney, Belinda, ERP, DJ, Kaz, Kev, Jamie, Peggy, Laurie, Beth, Claire, Pommy, Mary, Marilyn, and, of course, Colby. You are the best!

–J. Parker

PART I

Learning About Qualitative Research

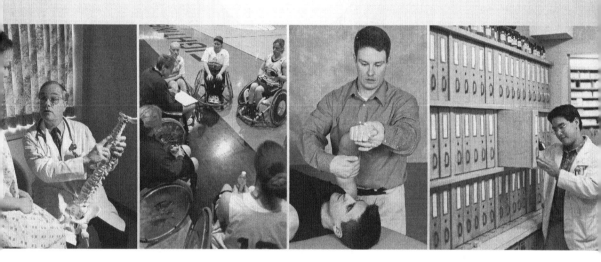

The journey begins with an introduction to the nature and structure of qualitative research. Chapter 1 specifies both the unique qualities of qualitative research and the systematic and scientific attributes that it shares with forms of research that are more traditional. Chapter 2 previews parts II and III with an outline of the structure of a qualitative study.

Guiding Questions

Consider the following questions before reading part I. They will guide your examination of each chapter.

1. What constitutes research and why is research important?
2. How is qualitative research defined?
3. What are the similarities and differences between qualitative and quantitative research?
4. What are the characteristics of qualitative research?
5. What are the components of a qualitative study?
6. How are the components of a qualitative study organized?
7. How would you succinctly and accurately record information from a qualitative study for future use?

The following figure illustrates the content, connections, and organizational configuration of part I.

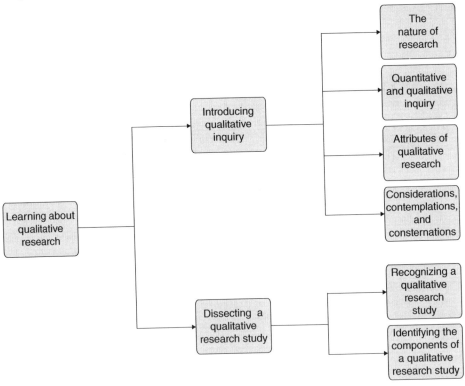

Introducing Qualitative Inquiry

Learning Objectives

Readers will be able to do the following:

1. Describe the nature of research.
2. Define qualitative inquiry.
3. Identify and describe the attributes of qualitative inquiry.
4. Compare qualitative and quantitative research.

The Nature of Research

Though this text focuses explicitly on qualitative inquiry, it must first define the term *research* to provide a context for the discussion. Research is viewed in many ways. Practically, it denotes the process of gathering information to solve a problem or answer a question (Booth, Colomb, & Williams, 2003).

We all face problems that intuitively engage us in the research process on a daily basis. For example, consider a subject who is interested in improving her cardiorespiratory fitness but has a preexisting knee injury that limits her activity level. Budgetary restrictions further complicate her problem. With these problems in mind, she might gather information on equipment options and prices from fitness facilities, retailers, and manufacturers' Web sites. She might also ask an athletic trainer or physical therapist about different forms of exercise. These responses to the problems make up an informal research process. A more sophisticated and formal research process is used for complex problems and for professional inquiry.

This broader definition explains the key tenets of research for professionals. Research is a systematic way of collecting and analyzing information to answer a specific question and to add to a discipline's knowledge base. It also is a way to systematically investigate a topic, phenomenon, issue, or problem of interest for greater understanding (Stringer, 2004).

The first tenet is the systematic nature of research. Researchers follow specific steps to solve problems, including collecting appropriate data, analyzing that information, and drawing reasonable conclusions from it. Scholars often question how the information gained from a study lends to their overall understanding of a topic. The second tenet is that research advances the understanding of a specific discipline or significantly relates to an area of study. Although these two tenets are appropriate for every form of research, the methods differ for quantitative and qualitative inquiry.

Quantitative and Qualitative Inquiry

Quantitative inquiry, also referred to as *traditional* or *conventional research* (Erlandson, Harris, Skipper, & Allan, 1993), is familiar to most students and practitioners. The term *quantitative* denotes measurement, and these types of studies represent meaning with numbers. For example, a quantitative analysis of the success of an exercise physiology course draws statistics from a questionnaire that equates level of satisfaction with numerical rankings (5 = very satisfied, 4 = somewhat satisfied, 3 = neutral, 2 = somewhat dissatisfied, 1 = very dissatisfied).

Many health care professionals use quantitative research to analyze numerical data, including heart rate (measured in pulses per minute), blood pressure (measured in mm/Hg), and blood-sodium level (measured in mmo/L). Quantitative researchers answer questions by identifying variables, measuring them, and examining how they relate to one another. When examining the relationship between exercise intensity and heart rate, researchers may ask whether a treatment or intervention causes a specific outcome or whether a cause-and-effect relationship exists between the two variables. In this example, exercise lowers blood pressure and long periods of exercise dilute blood-sodium levels if participants' intake of sodium and water does not match their output of sweat.

Although quantitative research is important and necessary, many aspects of professional and personal lives cannot be explained with numbers. Qualitative research is helpful in these instances. This alternative form of inquiry, although less familiar to health professionals, was commonly used in the 1920s and 1930s in large scientific disciplines like psychology. As the discipline of psychology expanded, the emphasis shifted to behaviorism and experimental design (Hayes, 1997). However, qualitative methods have regained popularity in recent years.

Qualitative inquiry is the new kid on the block in some disciplines, but it has gained acceptance in the last 10 years as a legitimate form of scholarship. Researchers in disciplines related to physical activity and health have used qualitative methods for more than 20 years to expand their methodological base and to broaden their understanding of human behavior (Harris, 1983). In fact, qualitative research methods are now extremely popular in the medical professions (Barbour, 2001).

Qualitative research has reached its current level of popularity and acceptance despite many scholarly debates about its value, legitimacy, and rigor (Paul, 2005). Many members of health and medical communities are skeptical about qualitative methods (Malterud, 2001). Devers and Frankel (2000) believe that quantitative researchers misunderstand the process, purpose, and products of qualitative research.

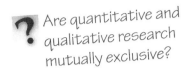 Are quantitative and qualitative research mutually exclusive?

No, some research problems lend themselves to investigation with both methodologies.

Attributes of Qualitative Research

Many unique attributes differentiate qualitative research from conventional, quantitative research. Numerous researchers have described it in terms of its characteristics (Merriam, 1998; Bogdan & Biklen, 2007), assumptions (Schram, 2006), and features (Silverman, 2000). This section identifies the unique qualities of qualitative research and compares and contrasts them with quantitative research.

- *Focuses on people.* Qualitative inquiry is extremely humanistic. Qualitative researchers are interested in how people perceive their experiences, what they believe about issues, and how their interactions with others influence these attitudes and values. These scholars study the concept of social construction, or the meaning people assign to their life situations based on their interactions with others (Berger & Luckmann, 1966). Most qualitative researchers believe that human perception of experience can rarely be measured and analyzed with numbers.

- *Uses textual data.* Because the meaning of human experiences cannot be represented by numbers alone, qualitative researchers interpret situations with personal descriptions and accounts. They conduct interviews and use their transcripts as data. Observations or documents may also be used as data. In each case, the information is collected and analyzed by a very sensitive instrument—the researcher. The careful researcher can comprehend complex situations and identify processes, perspectives, and perceptions that technical instruments might miss.

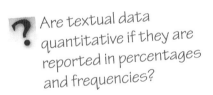 Are textual data quantitative if they are reported in percentages and frequencies?

No, some qualitative researchers use numbers to provide an overall picture of the data. The study's underlying attributes, research questions, and methods of data collection determine whether the research is qualitative or quantitative.

- *Discovery and exploration in natural settings.* Because the purpose of qualitative research is to better understand the human condition, another distinguishing attribute is discovery and exploration in natural environments. The laboratory environment for quantitative methods of human study, such as exercise science, is often more sterile and foreign than the environment in which people actually live and function. Qualitative researchers wish to understand the experiences of their participants in their natural settings, without manipulating or controlling the environment. For example, to fully understand the natural context, many qualitative researchers travel to schools to observe physical education classes; interview teachers, coaches, and students; and collect documents related to the study. This holistic approach deepens their understanding of human experience.

• *Interprets with inductive reasoning.* Qualitative research uses an interpretive process that relies on inductive analysis. In other words, researchers construct general findings from small pieces of specific information and then thematize them, or group them together in meaningful ways, to develop the results of a study. This process contrasts with deductive reasoning, in which general principles and information lead researchers to a specific conclusion. In essence, qualitative researchers form their conclusions over time as they collect and analyze data.

• *Systematic yet flexible.* The process of qualitative research is very deliberate and systematic. To ensure authenticity, researchers use specific tactics to design and plan the study, identify and select appropriate participants, and methodically collect and analyze data. However, qualitative research is also inherently flexible. It is often difficult for researchers to predict whom they will interview, which documents they will examine, or where they will conduct observations. Oldfather and West (1994) compared the improvisational aspect of qualitative research to jazz music because studies often take a different direction in response to new discoveries. As a study deepens and progresses, researchers may need to collect more data or adjust the project's time line in order to fully understand a complex situation.

• *Small sample size.* Because another goal of qualitative research is to gain insight, researchers strive more for a depth of understanding than a breadth of information

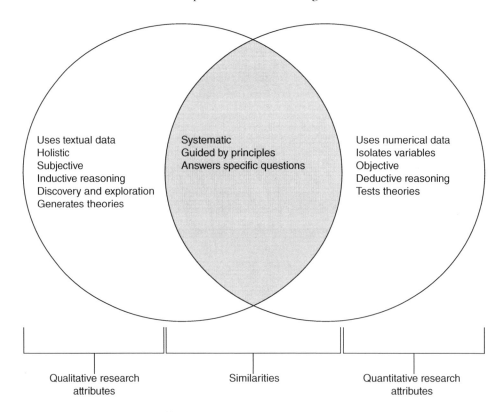

Figure 1.1 Similarities and differences between qualitative and quantitative research.

(Shank, 2006). They rarely seek to generalize their findings, so large numbers of participants are uncommon. Samples can range in size from one person, as in a case study or biography (Creswell, 2007), to 60 participants with other forms of qualitative research.

- *Provides rich descriptions.* Qualitative researchers provide detailed descriptions of the setting so readers can understand the participants' experiences and the meaning they assign to their situations and environments. They also enhance their reports with quotes that showcase the participants' voices and the essence of the study's findings.

- *Identifies data patterns.* Many qualitative researchers discover information about the participants' perceptions of their experiences that reveals commonalities. They group these emergent themes into categories that identify patterns of data.

- *Builds theories.* Conceptual model development is the process of constructing a theory, or a set of explanatory concepts, with advanced forms of data analysis (Merriam, 1998). Researchers theorize by interpreting data, and then not only explaining what has occurred, but also identifying possible reasons for the occurrence. Many researchers use visual models to illustrate their theoretical findings.

Figure 1.1 compares and contrasts qualitative and quantitative research. Remember that both forms of inquiry are equally valuable for addressing research questions.

Considerations, Contemplations, and Consternations

Qualitative research is not for the faint of heart. The conclusion of this first chapter presents aspects of qualitative research to consider and contemplate, as well as some qualities that may concern researchers.

Silverman (2000) suggests that many individuals gravitate towards qualitative inquiry to avoid statistical analysis, but soon find that the process is rigorous in its own right. Researchers should be aware of the attention to detail and level of organization required to accurately execute a qualitative study. Consider also that despite the recent success of qualitative research, many scientists are still resistant to this alternative form of inquiry. You must fully understand your method and why it is appropriate for your study so you are able to defend it when necessary.

The many forms of qualitative research can confuse novices. Each variation has a unique focus, method, and set of outcomes, but regardless of the final format, all qualitative research follows the basic inductive approach presented in this text.

Finally, two misconceptions plague many qualitative researchers. The first is that qualitative research is often not considered a form of scientific inquiry. The second is that the two forms of research are often viewed as mutually exclusive. That is, a study may employ either quantitative or qualitative research, but never both.

Is Qualitative Research Scientific or Not?

Many scholars who are more familiar with quantitative methods dismiss qualitative research as unscientific because it is subjective. Namely, qualitative findings come from interpretations of experiences rather than from measurable outcomes. These critics of qualitative inquiry believe that the aim of research is to observe or measure a single reality (Munhall & Boyd, 1993).

However, other scholars view qualitative inquiry as a form of science (Parse, 2001) because its systematic approach is guided by distinct principles derived from the scientific method. Shank (2006) calls this process qualitative science and suggests that the search for meaning differentiates qualitative scientists from their quantitative counterparts.

Is One Method Better Than the Other?

These issues have led to a paradigm war among scientists, who argue that one form of research is more rigorous, meaningful, and appropriate for the disciplines of health and physical activity. However, we argue that quantitative and qualitative approaches need not compete at all. Both are significant, necessary, and valuable forms of inquiry that achieve different purposes and answer different questions. Both forms of research can be rigorous, if done correctly.

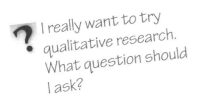 I really want to try qualitative research. What question should I ask?

The research question determines the type of inquiry, not the other way around. If you are truly interested in a phenomenon that requires methods of qualitative investigation, then qualitative research is for you! However, if your interest is best addressed through quantitative methods, you should stick with that form of research.

It is inappropriate to acknowledge a singular approach to research. All scholars have natural preferences for one form or the other, but a study's purpose and questions should drive the research method. Researchers must become familiar with both paradigms.

Summary

Research is the systematic process of collecting and analyzing information to answer specific questions. Qualitative research is a legitimate form of inquiry that allows scholars to gain insight and understanding about the human condition. Its key attributes include a humanistic orientation, a focus on discovery and exploration, and the use of inductive analysis. Qualitative researchers draw meaning from textual data rather than from numbers, and work with small groups of participants. Other attributes of qualitative research include rich descriptions, the emergence of data patterns, and the development of conceptual models. Although qualitative research has gained acceptance in many disciplines, it also has many critics. Both qualitative and quantitative forms of research are important and necessary in the disciplines of health and physical activity.

CONTINUING YOUR EDUCATIONAL JOURNEY

 Learning Through Activity

1. Use Web-based or textual resources to explain the difference between inductive and deductive logic. How is each form of analysis used in qualitative and quantitative research?

2. In addition to the ideas provided in this chapter, give another example of quantitative and qualitative data for the disciplines of health and physical activity.

3. Use a search index to locate five research articles that interest you. Using the basic research tenets as a lens, identify which of the articles are quantitative and which are qualitative in nature.

4. Reflect on this chapter and look ahead to the future chapters, then list any questions you have about how qualitative data is collected and analyzed.

 Checking Your Knowledge

1. Qualitative research uses large sample sizes to generalize research findings for the broader population.
   a. true
   b. false

2. Which term denotes the meaning that people assign to their interactions with others?
   a. humanistic development
   b. social development
   c. social construction
   d. humanistic orientation

3. Quantitative research uses methods of inductive analysis and interpretive processes.
   a. true
   b. false

4. Because qualitative research is exploratory in nature, it is both flexible and systematic.
   a. true
   b. false

5. Qualitative research uses which of the following?
   a. inductive analysis
   b. systematic, but flexible, methods
   c. humanistic approach
   d. discovery in natural settings
   e. all of the above

6. Which of the following is a principle related to the general research process?

    a. Methodical procedures should be used.

    b. Advanced understanding is an outcome of the research process.

    c. Only measurable data are meaningful.

    d. a and b

 ## Thinking About It

1. A colleague states that because qualitative inquiry fails to identify cause-and-effect relationships between variables, it is not a valuable form of research for athletic training or physical activity. What is your initial reaction to a statement like this? Explain.

2. Think of the personal interactions you have in your professional life and identify a question that would best be answered with qualitative data.

 ## Making a Stretch

Many writings exist that will help you stretch your mind and further explore the nature of qualitative research. Examining past arguments about qualitative methods that have surfaced in the physical activity and health professions will serve you well, providing both an overview of the research form and a context for its current professional position.

Bain, L.L. (1989). Interpretive and critical research in sport and physical education. *Research Quarterly for Exercise and Sport, 60*(1), 21-24.

Locke, L.F. (1989). Qualitative research as a form of scientific inquiry in sport and physical education. *Research Quarterly for Exercise and Sport, 60*(1), 1-20.

Martens, R. (1987). Science, knowledge, and sport psychology. *The Sport Psychologist, 1*(1), 29-55.

Sage, G.H. (1989). A commentary on qualitative research as a form of scientific inquiry in sport and physical education. *Research Quarterly for Exercise and Sport, 60*(1), 25-29.

Schutz, R.W. (1989). Qualitative research: Comments and controversies. *Research Quarterly for Exercise and Sport, 60*(1), 30-35.

Siedentop, D. (1989). Do the lockers really smell? *Research Quarterly for Exercise and Sport, 60*(1), 36-41.

Dissecting a Qualitative Research Study

Learning Objectives

Readers will be able to do the following:

1. Recognize a qualitative research study.
2. Identify the study's components.
3. Deconstruct the study.
4. Record the study's information accurately and concisely.

Recognizing a Qualitative Research Study

Before conducting a qualitative study, you must be able to recognize qualitative research when you see it. Figure 2.1 illustrates the components of a qualitative research study and provides practical advice for confirming the nature of a given study.

The clues are primarily embedded in the introduction and method sections. First, look through the introduction and abstract, if one is provided, for the purpose statement or research questions. Statements and research questions about a search for meaning, insight, understanding, experiences, or perceptions often indicate a qualitative approach. Second, go to the methods section and look for clues in the portions outlining participants and data collection. How does the study name the people involved? Qualitative researchers refer to their interviewees as *participants* rather than *subjects.* How many people are involved in the study? Qualitative studies deliberately select relatively small numbers of participants compared to hundreds or thousands of subjects in some quantitative research studies. Finally, how is information collected? In qualitative research, the researcher serves as the instrument for data collection. Look for observations and interviews conducted by the researcher and references to transcripts, field notes, or both.

Selected text on pages 13-16, 18-20, and 22-25 reprinted, by permission, from W.A. Pitney, 2002, "The professional socialization of certified athletic trainers in high school setting: A grounded theory investigation," *Journal of Athletic Training* 37(3): 286-292.

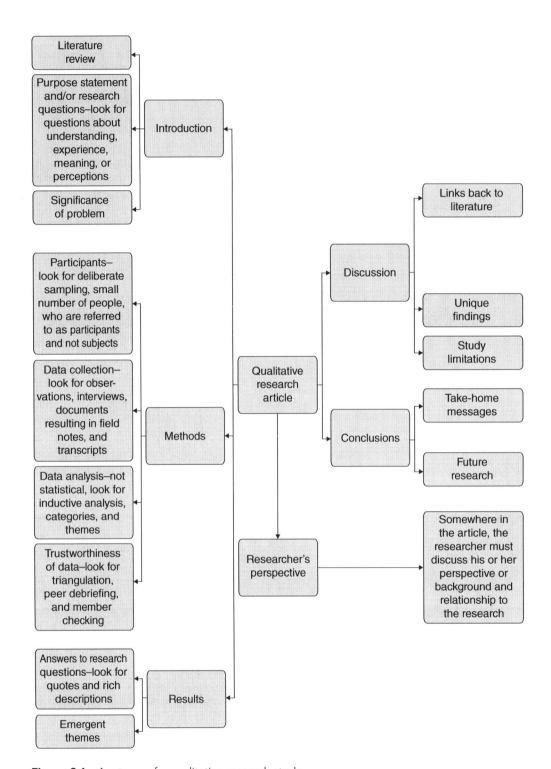

Figure 2.1 Anatomy of a qualitative research study.

If you have confirmed the presence of all the preceding criteria, congratulations, you have located a qualitative research study! If you have found some of the criteria to be absent, you have located a different kind of research study that may be valuable later. Other forms of research are important and have made vital contributions to the knowledge base of qualitative studies.

Identifying the Components of a Qualitative Research Study

Once you have found a qualitative study, you may begin identifying its components and recording the information. The remainder of the chapter outlines the components of a single qualitative study. Each section features a shaded box with a concrete example from the published study in appendix A and a prompt that will help you apply the information to your own study. Please note that we do not recommend copying large chunks of text from articles into your report. The intent of the quoted excerpts is to direct you to the matching section of the article and to help you succinctly summarize the information presented. Consider highlighting the critical points as you read through the shaded boxes and checking those points against the information presented at the end of each section.

Your first question should be, "What is the full citation of this study?" When you have located a qualitative research article, you must accurately and completely record the citation. Use the most recent reference manual for the citation format preferred by your advisor, committee, and academic discipline. This step is critical because you will need to refer back to the study later on, and it is very frustrating to work with an incomplete citation. Here is the correct American Psychological Association (APA) format for the study deconstructed in this chapter:

> Pitney, W.A. (2002). The professional socialization of certified athletic trainers in high school settings: A grounded theory investigation. *Journal of Athletic Training, 37,* 286-292.

I know I should record the citation in an appropriate format, but it just seems so cumbersome. Can I develop my own format?

Bad idea! Shortcuts are appealing, but they lead to errors. The more you practice using a specific formatting style, the easier it will become. Standardized references are crucial for accessing articles in the future.

Introduction

The introduction sets the scene for readers and indicate the study's direction. It usually contains three critical pieces of information: a short literature review, the purpose of the study and/or the research questions, and a statement about the study's significance.

Literature Review

The introduction usually begins with a short review of the articles that informed the study. It should also identify the study's conceptual framework. Read carefully for theories and articles that are critical to the author's perspective. The study deconstructed here is grounded in the literature of professional socialization:

Professional socialization is a process by which individuals learn the knowledge, skills, values, roles, and attitudes associated with their professional responsibilities.[1] Socialization is considered to be a key component of professional preparation and continued development in health and allied medical disciplines[2,3] and has been investigated in medical education,[4,5] nursing,[6,7] occupational therapy,[8] and physical therapy.[9]

Professional socialization is typically exemplified as a 2-part developmental process that includes experiences before entering a work setting (anticipatory socialization) and experiences after entering a work setting (organizational socialization).[10] The first process, anticipatory socialization, refers to experiences such as one's formal training as an undergraduate or graduate student, background as an employee in another setting such as an Emergency Medical Technician, or prior experience as a volunteer with an organization such as the American Red Cross. Organizational socialization refers to experiences such as in-service education and mentoring. The organizational socialization phase of professional socialization can be divided into 2 parts: (1) a period of induction, and (2) role continuance.[10]. Induction experiences take many forms. For example, induction processes can be either very formalized (ie, requiring employees to attend specific orientations or instructional sessions) or very informal (ie, no orientation). Additionally, induction processes may be either sequential, requiring specific skills to be learned at specific times during the initial periods of a job, or random, having no time frame for the development of various skills within the organization. Role continuance, on the other hand, focuses on adjusting to the organizational demands over time and continually learning the nuances of a given role and developing professionally…. Organizational socialization relates to how individuals adapt to their new roles and learn about what is acceptable practice in dealing with the demands of their work. For example, the organizational socialization can be very structured, such as having an athletic director orient a new employee in a very systematic manner, or this process can be unstructured, leaving the employee to ask questions of other employees as various situations arise. Understanding the organizational aspects of professional socialization allows the discovery of the necessary aspects of professional development in a work setting and can serve to improve both athletic training education and continuing education strategies.

Identifying and correctly citing the critical articles may help you discover additional articles in the future. The first place you should look for further information on a topic is the article's reference list.

I have located what I believe to be a qualitative research article on the Internet. However, I cannot find any references to critical articles. In fact, the authors hardly mention previous studies. Why is this?

Congratulations, you have identified an important research problem. Unless the article is from a scholarly journal, it could have been posted on the Internet by anyone and may be more of an opinion piece than a systematic research study. You may also have retrieved an abstract or research summary. If so, look through the full article for references.

Here are the critical articles identified by Pitney (2002):

Clark, P.G. (1997). Values in health care professional socialization: Implications for geriatric education in interdisciplinary teamwork. *Gerontologist, 37,* 441-451.

Teirney, W.G. & Rhodes, R.A. (1993). Faculty socialization as a cultural process: A mirror of institutional commitment. *ASHE-ERIC Higher Education Report No. 93-6.* Washington DC: George Washington University School of Education and Human Development.

Purpose Statement and Research Questions

The literature review should lead you to the research questions or to the statement of the purpose. Authors often use the literature to summarize what is already known on the subject and to explain how the research will contribute to the discipline's knowledge base. At this point, you should be able to identify the study's path. What is the purpose of this study? What specific questions does it intend to answer?

The purpose of this study, therefore, was to explore the professional socialization of certified athletic trainers (ATCs) in the high school setting in order to gain insight and understanding into how they initially learned and continued to learn their professional responsibilities in an organizational setting.

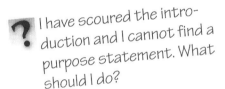
I have scoured the introduction and I cannot find a purpose statement. What should I do?

It is the responsibility of the authors to articulate the purpose of the study, but sometimes they fail to make it clear. Words like *goals, objectives,* or *intent* will help you identify the purpose statement.

Study's Significance

Research studies must contribute to and extend the current knowledge base. Ask yourself why this study is significant. What will it tell you that you don't already know? An explanation of the study's importance should closely follow the purpose statement and research questions in the introduction.

While athletic training has given a great deal of attention to the anticipatory socialization by way of the professional preparation process, there is a paucity of research related to organizational socialization.

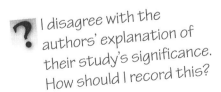
I disagree with the authors' explanation of their study's significance. How should I record this?

At this point, you should not pass judgment on what you read, but merely record the facts. Chapter 10 explains how to evaluate the content of an article.

In this example, the author identifies a lack of research about organizational socialization in athletic training. He also explains how knowledge about this subject may enhance future athletic training and education.

> Understanding the organizational aspects of professional socialization allows the discovery of the necessary aspects of professional development in a work setting and can serve to improve both athletic training education and continuing education strategies.

The preceding text further explains the significance of the study and begins to identify how it will contribute to the discipline. The schematic in figure 2.2 illustrates the important points from this study's introduction. Some researchers prefer to record information with graphics because they are easy to use and enable memorization.

Methods

The methods section should provide information on the participants and the data. It should outline how the researchers protected the participants, collected and analyzed data, and ensured trustworthiness of the data. You may also learn about the researcher's perspective. This is important because the qualitative researcher serves as the instrument that collects data, and you should understand his or her perspectives, experiences, and connection to the investigation.

Researcher Perspective

In the following study, Pitney describes his perspective before he introduces the participants. This technique gives readers a lens through which to view the entire study. Ask yourself how the authors describe their beliefs about and background in the research topic.

> With qualitative methods, the researcher is the primary "instrument" for data collection and analysis, and extreme sensitivity is given to the nature and perspectives of human participants. A researcher's perspective, however, can shape the analysis and interpretation of the qualitative data. My perspectives about the high school setting were shaped in 2 ways. First, I was formerly employed as a clinical high school ATC and interacted with coaches, athletic directors, and athletes and their parents. Second, at the time of data collection and analysis, I was a faculty member in a Commission on Accreditation of Allied Health Education Programs (CAAHEP)-accredited program that used several high school sites as clinical education experiences for the athletic training students.[14]

Protection of Participants' Rights

You must have a thorough understanding of the participants in a qualitative research study. The methods section should provide detailed demographics and background information on the participants. It should also specify how their rights were protected. This piece of information is often short and can sometimes be reduced to a single sentence indicating that the study was approved by the appropriate Institutional Review

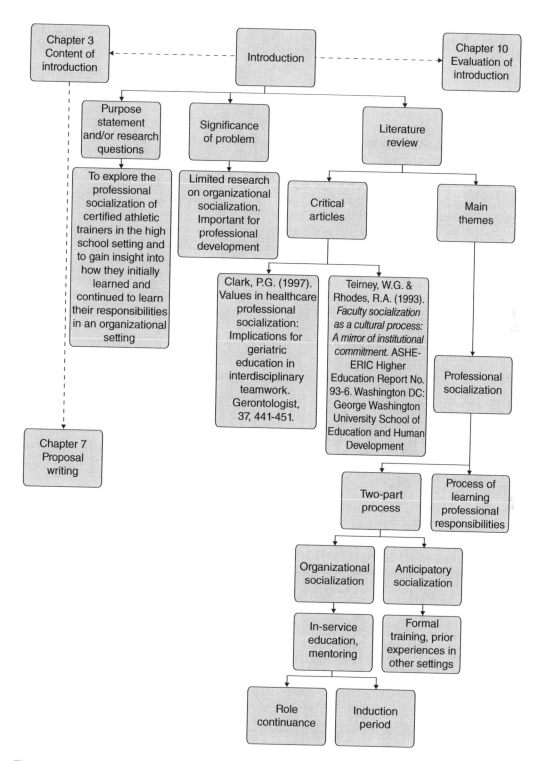

Figure 2.2 Schematic representation of the introduction.

Board (IRB). The breadth of information about the study's participants corresponds to the amount of risk they assumed. Ask yourself who the study's participants were and how their rights were protected.

In qualitative research, the protection of the participants' anonymity is paramount. Therefore, audiotape recordings of interviews were transcribed and labeled with pseudonyms that are used at various locations in the manuscript. Moreover, at the completion of the study, the audiotapes were destroyed, but the transcripts were maintained using the established pseudonyms. Before data collection, appropriate institutional review board approval was received.

A total of 15 individuals (12 ATCs currently practicing at the high school level, 2 current high school athletic directors who are also ATCs, and 1 former high school ATC) participated in the study. The average number of years in their current position for the 12 currently practicing ATCs was 10.16 ± 7.44, with a range of 2 to 25 years. The 2 athletic directors averaged 5.5 years of experience, and the former high school ATC had worked in that particular setting for 1 year before entering graduate school. The athletic directors and former high school ATC were included in order to cross-reference, or triangulate, the perspectives of the currently practicing ATCs. Six participants were women and 9 were men. The participants represented 3 different midwestern states. Participants were initially purposefully selected: that is, I recruited volunteers whom I knew and who agreed to interviews. I then asked these individuals for suggestions of other ATCs who might be willing to participate. The remaining individuals were contacted via either e-mail or phone and agreed to interviews. Before the interviews, participants were required to review and sign an informed consent form.

In this study, Pitney (2002) uses a table to clearly identify the participants' demographic information and to support the details outlined in the text. If the study you are reading includes a similar table, we suggest that you attach a photocopy to your record sheet.

Data Collection

The author is responsible for providing detailed information about the process of data collection. Your role as the reader is to locate and record this important information. Ask yourself what methods were used for data collection. How were they conducted? Specifically, how long did they take? What was the sequence of events?

Data were collected using semistructured interviews. Each interview incorporated several key questions or open-ended statements, including the following:

1. Describe your first few years of being an ATC at the high school level.
2. How did you learn your role and professional responsibilities at the high school level?
3. What has been your greatest challenge at the high school level, and how did you learn to deal with it?
4. What do you like best, or what are the good things about your current position?

5. What aspect of your job do you feel least satisfied by?

6. What is, or how would you describe, your professional mission?

7. What motivates you on a daily basis?

8. What advice might you give to an ATC just about to enter the high school setting for the first time?

Because both athletic directors were former ATCs practicing in the high school setting, they were asked to reflect on their experiences as an ATC by answering questions 1 through 4 and question 8. Additionally, they were asked to describe the priorities of the athletic department, the role of ATCs in the high school setting, and the challenges that ATCs face in the high school setting. The interviews ranged in length from approximately 35 to 105 minutes. Eight interviews were conducted by phone, and 7 interviews were conducted in person, based on feasibility and availability. Participants gave advanced written and verbal consent to tape record the interviews. The tape-recorded interviews were then transcribed and analyzed using a grounded theory approach. Data were collected until theoretic saturation was achieved.[15]

This segment of the article fully describes how the author collected and managed the data. Researchers interested in conducting a similar study would find enough information to guide their process.

Data Analysis

Next, the authors must explain how they analyzed the data. Often, they refer back to the theoretical framework to explain their procedural choices. At this point, you don't need to understand the actual process of data analysis, but you should be able to recognize and record the methods. Ask yourself how the authors analyzed the data. Which specific process did they use? What were the outcomes, such as themes or categories, of that process?

The grounded theory approach, as discussed by Glaser and Strauss,[16] is helpful for generating theory (a set of explanatory concepts) based on the data collected. I specifically used open, axial, and selective coding procedures documented by Strauss and Corbin.[15] Raw data were analyzed inductively, and incidents or experiences related to the phenomenon under investigation were identified and labeled as a particular concept. This type of coding strategy is described as "creating tags," and the purpose is to produce a set of concepts that represents the information obtained in the interview.[17] Identifying these concepts and placing them into like categories based on their content completed the formal open-coding process. Relating categories with any subcategories that might exist and examining how one category related to another completed the axial-coding process. Selective coding involved integrating the categories into a larger theoretic scheme and organizing the categories around a central explanatory concept, specifically, the proposition that informal learning processes were critical to the successful professional socialization of ATCs in the high school setting.

Ensures Trustworthiness of Data

This section is unique to qualitative research studies and is crucial because it addresses issues of believability and credibility. The question of trustworthiness of data can often be answered with a single sentence that lists a specific technique, but it should appear in every qualitative research study. Ask yourself how the author addressed issues of trustworthiness in this study.

> Several techniques were employed in order to establish trustworthiness of the data collection and analysis, including peer debriefing, data-source triangulation, and member checks. A peer debriefing was completed by having an athletic trainer with a formal education in qualitative methods (a minimum of 3 qualitative research methodology courses at the doctoral level) review the documented concepts and thematic categories for relevance, consistency, and logic. Moreover, the reviewer examined the interview questions in each transcript to determine if they were "leading" in nature. The textual data from any questions identified as being leading were not included in the analysis. The reviewer was in agreement with the findings based on the purpose of the study and even suggested other concepts that would strengthen one particular category. Data-source triangulation, which is cross-checking perspectives,[18] was obtained by interviewing current high school athletic directors and a former high school ATC. Member checks were completed electronically by e-mailing the results to 5 participants and allowing them to comment on the thematic categories. Three individuals responded, agreed with the results, and had no further input, indicating no misinterpretation of the professional-socialization process that emerged from this study. On an informal basis, I also explained the results to 4 other participants, and they were in agreement with the findings.

Figure 2.3 schematically represents the methods section of Pitney's study (2002). Note that the information is condensed to focus on the key points. If you are curious about additional details, reread specific sections of the study (see appendix A).

Results

The results section describes the themes or categories that emerged from the data. They should be supported by direct quotes from participants. The authors should also attempt to summarize their findings with a concise statement. When reading the results section, you do not need to record all the participant quotes. Ask yourself what themes emerged from the data. How do the authors make sense of those themes?

> The concepts identified during the open coding were organized into 2 categories that gave insight into the professional socialization process: (1) an informal induction process, aspects of organizational learning, and (2) creating networks for learning.

Here, you would record the two themes that emerged from the analysis. The author follows the explanation of the themes with supporting information and quotes from the participants. This section also includes a summary statement that concisely captures the results:

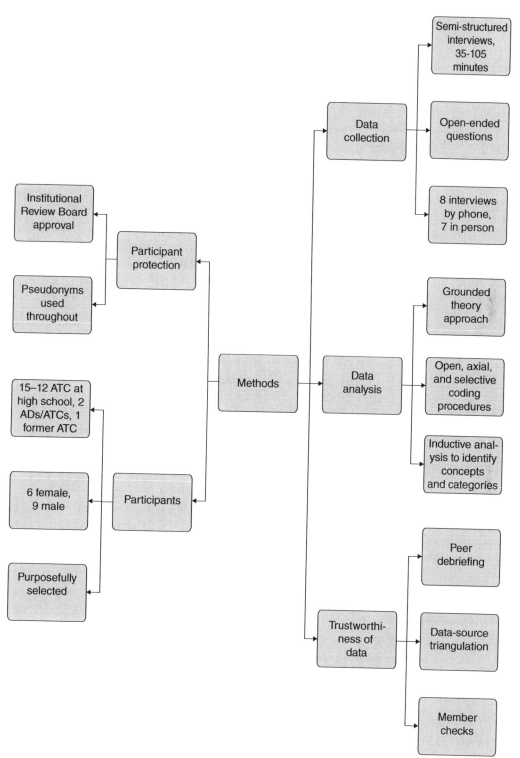

Figure 2.3 Schematic representation of the methods.

Given the nature of the 2 thematic categories, the resultant proposition is that informal learning procedures are critical to the professional-socialization process for ATCs working in the high school setting. The results of this study suggest that learning through informal means, such as collegial networks, organizational peers, and trial and error, are necessary elements for navigating the high school work setting and being socialized into the ATC role.

Again, the schematic in figure 2.4 illustrates how the results may be summarized and recorded. If the article includes quotes that are particularly powerful, you may wish to note their location for easy reference.

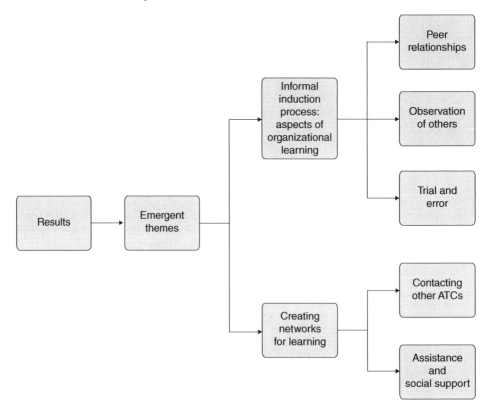

Figure 2.4 Schematic representation of the results.

Discussion

The author should pull everything together in this section, discussing clear links to the literature, details of any unique findings, and the study's limitations.

Makes Connections With Literature

The authors should revisit the articles from the literature review in the discussion section to situate their findings within the current base of knowledge. Specifically, they should explain how their study supports existing knowledge and provide a context for new insights. When recording this information, ask yourself how the study's findings supported the literature from the review. What unique contributions did the study make?

The socialization literature[6,10,19-22] concludes that the initial entry into an organizational setting is a period of adjustment for many professionals. This study supports these findings, as participants suggested that there was an initial adjustment when entering the high school setting from their previous setting (undergraduate program, master's program, or previous job). Additionally, many participants learned through informal means such as trial and error and by observing others in the organizational setting. Based on a socialization study by Ostroff and Kozlowski,[23] this is not unusual, as many individuals frequently rely primarily on observation of others and trial and error to acquire their information in an organization. Thus, informal learning plays a critical role in the professional-socialization process.

In this excerpt, the author identifies the role of informal learning in the socialization process as a unique contribution.

Study Limitations

Every study has limitations, and it is the authors' responsibility to clearly articulate them. Let this section about the strengths and weaknesses of research inform your own research practice. As you read, keep track of the study's limitations.

Most of the participants in this study were practicing at schools located in metropolitan areas as designated by the US Department of Commerce, Bureau of the Census, yet none were located in an inner city school or extremely rural setting. The propositions resulting from this grounded theory, therefore, may not be transferable to the inner city or rural school context.

Figure 2.5 schematically represents the discussion section from Pitney's study (2002). Notice that the details have again been summarized with key points that should jog your memory during subsequent readings.

Conclusions

The conclusion should articulate in several short paragraphs the relationship between the results and the research questions. Did the results answer the research questions? The authors should also concisely condense the entire study into two or three main points. Finally, they should indicate a direction for future research on this particular topic. As you read, record the key points and ask yourself how they might influence future practice.

The organizational aspects of the professional-socialization process among high school ATCs are principally informal in nature. As such, the ATCs relied on informal learning strategies during their period of induction. To facilitate their continued development, informal

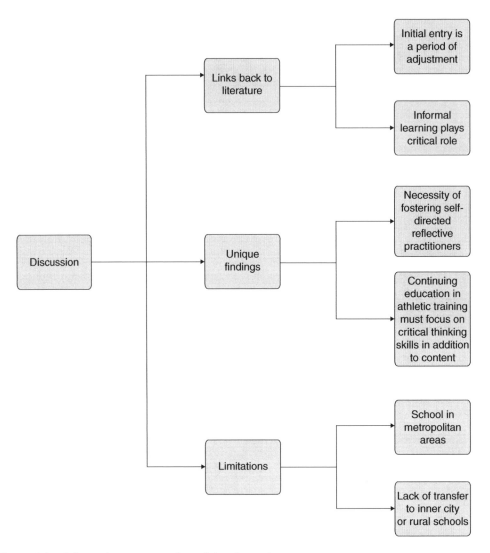

Figure 2.5 Schematic representation of the discussion.

learning networks that largely comprised colleagues outside of the organizational setting were created. To ensure that individuals effectively learn through informal means, preservice athletic training education programs would be well advised to foster the development of reflective practitioners who think critically and are self-directed and self-evaluative. This can be accomplished by using such educational strategies as reflective journals, learning plans, and independent projects. Continuing professional educators should also attempt to foster self-evaluation, critical reflection, and critical-thinking ability. Continuing professional educators can accomplish this by employing strategies such as verbal reasoning and problem solving and consciously raising questions and giving clinicians an opportunity to discuss their thought processes in a non-threatening learning environment. Moreover, it has been argued that continuing education should be linked to practical problems.

Future Research

This section is important for readers seeking to formulate their own research on a given topic. It may help you find a research niche or identify possible collaborators. What recommendations for future research does the author make?

> Because informal learning is highly contextual, multiple influences have the propensity to shape the extent to which informal learning is successful, including cultural factors, career structure, technology, and learning needs.[35] As such, future studies could investigate exactly how these factors influence informal learning and role socialization. Additionally, because informal learning can be inhibited in many ways, it may be helpful to examine the environmental inhibitors (ie, job demands) of informal learning in various athletic training settings.

Figure 2.6 schematically represents the conclusions of Pitney's article (2002).

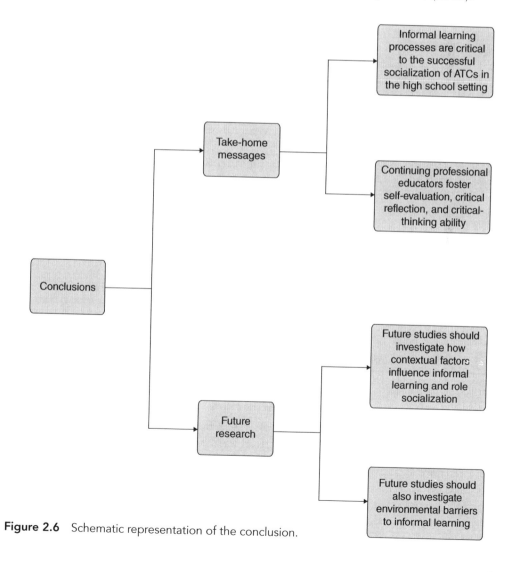

Figure 2.6   Schematic representation of the conclusion.

Summary

This chapter outlines the components of a qualitative research study, including a brief introduction to the content of each section. To best understand qualitative research, you should summarize and record relevant information as you read. The following 15 questions will help you summarize a study. We suggest that you answer these questions on a single sheet of paper, and then attach it to the study and keep it for your files. If you prefer to read directly from the computer screen, simply file the summary sheet alphabetically by author's name and save a few trees!

Summarizing a Qualitative Research Report

1. What is the full citation for this study?
2. What conceptual framework informs this study?
3. What critical articles do the authors use?
4. What is the purpose of this study and what specific questions does it set out to answer?
5. Why is this study significant? What might it tell me that I don't already know?
6. How do the authors describe their beliefs about and background in the research topic?
7. Who are the participants? How do the authors protect their rights?
8. What data collection methods are used and how was each one conducted (i.e., length of time, sequence of events, and so on)?
9. How do the authors analyze the data? Which specific process do they use and what are the outcomes (themes, categories, etc.) of that process?
10. How do the authors address issues of trustworthiness?
11. What themes emerge from the data? How do the authors make sense of them?
12. How do the findings of this study support the literature? What unique contributions does this study make?
13. What are the limitations of this study?
14. What are the key points of the study and what are the critical "take home" messages? How might these influence future practice?
15. What recommendations for future research do the authors make?

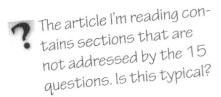 The article I'm reading contains sections that are not addressed by the 15 questions. Is this typical?

The 15 questions are designed to help summarize the components of a typical qualitative study. The questions may not correspond directly to all sections you may find in an article. Make sure you record all the information you need.

CONTINUING YOUR EDUCATIONAL JOURNEY

 ## Learning Through Activity

1. Locate five articles on a topic of your choice using an electronic search engine. Use the criteria from the beginning of this chapter to determine which articles include qualitative research. Print out or save one of the qualitative articles.

2. Read the article by Goodwin and Compton (2004) in appendix C and write a summary using the 15 questions provided in this chapter.

3. Repeat the summary process with the article you printed out from your electronic search. The more you read, the better you will become at recording!

 ## Checking Your Knowledge

1. Which section of an article provides critical information that determines whether a study is qualitative in nature?
   a. introduction and conclusions
   b. methods and results
   c. introduction and methods
   d. discussion and conclusions

2. Qualitative studies usually use large numbers of participants.
   a. true
   b. false

3. One of the best ways to record information from a qualitative research study is to reproduce large quantities of the original text.
   a. true
   b. false

4. The initial statement of a research problem is located in the article's introduction section.
   a. true
   b. false

 ## Thinking About It

1. As you begin the process of qualitative research, you will read many published articles and research reports. Knowing yourself as you do, how can you best organize your reviews, data, and article summaries for easy access at a later date?

2. You have recently read a study and are extremely interested in the topic. You want to know more about it but you're having trouble locating relevant information. How can the study you have just read inform your search?

 ## Making a Stretch

This book will expand your knowledge on identifying and recording information from qualitative research articles.

Locke, L.F., Silverman, S.J., & Spirduso, W.W. (1998). *Reading and understanding research*. Thousand Oaks, CA: Sage.

PART II

Conducting a Qualitative Research Study

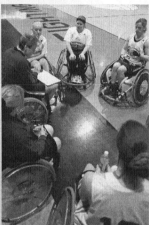

Chapter 3 opens part II with an overview of conceptualizing and planning a qualitative research study. Specifically, this chapter explains how to write an introduction containing a review of related literature, background information, a purpose statement, and research questions that inform the study's methods. Chapter 4 focuses on specific methodological strategies, including various forms of data collection, sampling, and data analysis. Chapter 5 explains the importance of ensuring trustworthiness in a qualitative research study, and chapter 6 addresses ethical issues connected to the research process.

Guiding Questions

Consider the following questions before reading part II. They will guide your examination of each chapter.

1. What is the critical link between a study's introduction and its methods?
2. What role does reviewing the literature play in the development of the introduction?
3. What does the term theoretical framework mean? How does it frame a study?
4. What are common methods of collecting and analyzing data?
5. What forms of data exist?
6. What are the eight steps for analyzing qualitative data?
7. What does trustworthiness mean in the context of qualitative research?
8. What does the term informed consent mean? What role does it play in qualitative research?
9. What steps are required for the protection of a study's participants?

The following figure illustrates the content, connections, and organizational configuration of part II.

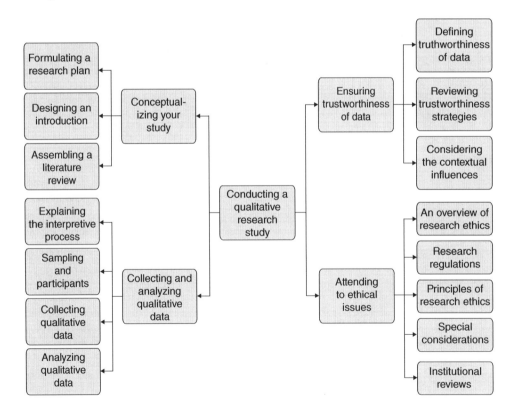

Conceptualizing Your Study

Formulating a Research Plan

Because qualitative research is flexible and adaptive, allowing scholars to change their data-collection tactics throughout the process, many critics believe that qualitative researchers do not plan sufficiently. However, qualitative researchers follow a planning process similar to that of any other type of study. They write proposals, follow principles, and identify strategies. In presenting this first chapter of part II, Conducting a Qualitative Research Study, we will discuss two important aspects of formulating a research plan: the introduction and literature review. We present these two components in a linear fashion but, in fact, you cannot formulate one component without involving the other to some extent. We will discuss this interactive process more later in the chapter, but for now we will first discuss the purpose and function of an introduction and provide tips for its creation. We will follow this by explaining the role of the literature review and offer sound advice on how best to accomplish a literature review and use it to inform your introduction. The actual proposal writing will be addressed in chapter 7.

Designing an Introduction

The introduction eases readers into your topic and helps them understand why you believe this particular study is necessary. It also places your study in a broader context. Here, you articulate your position and share critical research questions with readers so they understand what you hope to learn. Many researchers such as Marshall and Rossman (1999) have compared a study's introduction to a funnel that opens with a wide description and then gradually narrows into a tightly focused stream of information traveling in a clear and purposeful direction. Figure 3.1 portrays this conceptualization.

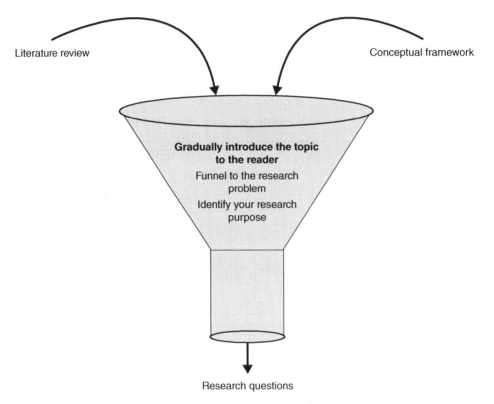

Figure 3.1 Use the funnel approach to organize your introduction.

In your introduction, you must begin by giving your readers some general background information. The first paragraphs should orient readers to the topic and give them a frame of reference for the study. Mensch and Ennis' study, "Pedagogic Strategies Perceived to Enhance Student Learning in Athletic Training Education," begins with the following paragraph:

> Since its inception in 1950, the National Athletic Trainers Association (NATA) has striven to enhance athletic trainers' knowledge and skills by improving educational experiences within its education programs. The transition of athletic training education from an internship program to a Commission on Accreditation of Allied Health Education Programs (CAAHEP) accredited competency-based program

has helped to standardize athletic training education and improve its consistency with professional preparation in other allied health disciplines. Instructors in both clinical and classroom settings are expected to integrate a common body of athletic training competencies across a wide range of student educational experiences. (2002, p. S199)

These authors set the stage for the study, eased readers into the material, and then ended with a statement outlining the multifaceted nature of the topic.

Mensch and Ennis next explain the study's conceptual framework with three key theories, including self-determination theory (SDT) and self-efficacy theory. The second paragraph opens with a sentence that funnels the reader's attention toward pedagogic strategies and student learning: "An essential component of good educational practices found consistently in research on teaching and learning is the integration of achievement motivation constructs into student learning" (p. S199).

Finally, the authors further narrow the focus of the text to a meaningful purpose by stating what we have learned from athletic training educational research, then argue this point: "The use of qualitative and descriptive research in athletic training education can benefit both students and instructors by providing a more meaningful understanding of educational practices" (p. S200). The funnel terminates in the following purpose statement:

Our purpose was to investigate students' educational experiences in CAAHEP-accredited athletic training education programs. Specifically our focus was to determine to what extent pedagogic strategies were reflected in students' perceptions of their learning experiences in CAAHEP-accredited athletic training programs, instructors' perceptions of their teaching in CAAHEP-accredited programs, and CAAHEP-accredited athletic training course syllabi. (p. S200)

Presenting a Conceptual Framework

The introduction should contain a conceptual framework, also called a conceptual context (Maxwell, 1996) or a theoretical framework (Merriam, 1998), that explains the key theories or concepts that support your position. This portion situates your study in the context of existing literature and theory. Silverman explains the critical relationship between study and theory: "[w]ithout theory, research is impossibly narrow. Without research, theory is mere armchair contemplation" (2000, p. 86). Merriam further emphasizes the importance of building the conceptual framework, which "draw[s] upon the concepts, terms, definitions, models, and theories of a particular literature base and disciplinary orientation" (1998, p. 46).

? What should I do if I don't have a conceptual framework for my study?

Ask yourself why. If your study is unique (which, to be honest, is rare at the master's level and a little more likely at the doctoral level) then you may have to stretch to include other bodies of literature to support your investigation. Otherwise, re-examine your search terms and get back to the library!

Consider the conceptual framework of Podlog and Eklund's 2006 study of competitive athletes' return to sport after serious injury. These authors effectively framed their study with Self Determination Theory (SDT), suggesting that the success of athletes' return to sports after an injury depends on whether the context addresses and meets their psychological needs. This example makes clear that the introduction draws from a review of the literature, and a review of the literature helps to identify the conceptual framework of a study. Use theories to frame your study even if they don't have specific names. For example, in a longitudinal study published in *Qualitative Health Research*, Thompson, Humbert, and Mirwald (2003) investigated how participants' level of physical activity during childhood and adolescence influenced their perceptions of exercise as adults. The following excerpt shows how they framed their study:

> Physical activity is promoted in children because it is thought that physically active children become physically active adults (Pate et al., 1999). In other words, it is believed that physical activity tracks from childhood to adulthood. Childhood is also considered the best time to socialize children into physically active lifestyles because it is the time when the attitudes and skills develop that are regarded as important for regular adult physical activity (Telama, Yang, Laakso, & Viikari, 1997). One explanation for this belief is that the experiences acquired and skills learned in childhood physical activities facilitate similar adult physical activities as well as the possibility of adopting new forms of physical activity/sport in adult years (Telama et al., 1997). As such, early physical activity experiences have been reported as important factors for predicting adult physical activity (Engstrom, 1991). (p. 359)

In this example, the authors place their research in the context of four studies. Again, the concepts, theories, and models from the literature are used to develop the framework.

Researchers sometimes have unique ideas for studies that are based more on personal experience than existing literature. Although this practice is acceptable, Maxwell (1996) points out that many scholars view frameworks grounded in experience as biased. Avoid bias by looking for literature that supports your position.

Explaining the Research Problem

The problem statement identifies the primary concern or issue that warrants investigation. According to Creswell, a research problem "is the issue that exists in the literature, in theory, or in practice that leads to a need for the study" (2003, p. 80). The length of the problem statement depends on the complexity of the issue and may range from a couple of sentences to a paragraph in length.

Regardless of its length, the statement must make the problem clear to the readers. Researchers often articulate a problem too quickly, assuming that their understanding of a phenomenon is common knowledge. Without a broader conceptual context and scaffolding of concepts, however, it is difficult for readers to make connections and understand the problem being researched.

Articulating the Purpose Statement

If an introduction is well written and logically leads readers to a clear problem statement, they will likely understand the study's aim before reading the purpose statement. Even so, authors must clearly state their intentions with a purpose statement that outlines their plans (Creswell, 2003; Marshall & Rossman, 1999).

A purpose statement can take many forms, though it often consists of one or two sentences (Marshall & Rossman, 1999) that identify the study's central idea (Creswell, 2003). Take the following steps to develop a purpose statement:

1. Draw the reader's attention to the statement.
2. State what you plan to do in the study.
3. Articulate a focus that directly relates to the problem statement and introductory material.

Drawing attention to the purpose statement is simple. Researchers often use one of the following phrases:

- "The purpose of this study is to . . ."
- "The objective of this study is to . . ."
- "Our study will seek to examine . . ."
- "The intent of our study is to . . ."

Here are some examples of complete purpose statements from published studies. Note that they are written in past tense because the studies have been completed. When developing a proposal, use the future tense to state your intent.

- *Example 1.* "The purpose of this investigation was to examine competitive athlete's experiences in returning to sport following a serious injury" (Podlog & Eklund, 2006, p. 46).
- *Example 2.* "The purpose of our study was to conduct an in-depth examination of male National Collegiate Athletic Association (NCAA) Division I ATCs who have been identified as leaders in their field based on longevity and contributions to the field of athletic training" (Malasarn, Bloom, & Crumpton, 2002, p. 56).

Developing Research Questions

Once you have funneled readers through your introductory paragraphs, problem statement, and purpose statement, you should provide your research questions. These serve as the tip of the funnel, or the culmination of all the content you have presented in your introductory chapter. Don't confuse research questions with questions you might ask during an interview to collect data. Research questions are large conceptual questions you hope to answer after your study is completed.

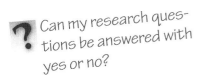 Can my research questions be answered with yes or no?

No! Rephrase your research questions for deeper exploration. Appropriate opening words include *what, how,* and *to what extent.* Avoid *why* questions because they contain an inherent assumption of definitive truth.

Some scholars choose not to conceptualize their studies with research questions. However, we have observed that most advisors want to hear about the research questions when students approach them about a study. We believe it is helpful to provide research questions when initially designing a study. Combined with your purpose statement, they will guide your methods by influencing which participants you select, what type of data you collect, and even what you ask during interviews.

What is the difference between a research question and an interview question?

Research questions are broad and ground the entire study. Interview questions are specific and serve as stepping stones toward the answers to research questions.

The following examples from previously cited studies show purpose statements followed by research questions:

- *Example 1*

The purpose of this investigation was to examine competitive athletes' experiences in returning to sport following a serious injury. . . . Two questions guided this study. First, what are the key psychosocial issues and processes involved in making the transition from injury rehabilitation to training and competition, and second, to what extent can SDT [self-determination theory] help to provide a greater theoretical understanding of the key issues and processes related to a return to sport following serious injury? (Podlog & Eklund, 2006, p. 46)

- *Example 2*

The purpose of our study was to conduct an in-depth examination of male National Collegiate Athletic Association (NCAA) Division I ATCs who have been identified as leaders in their field based on longevity and contributions to the field of athletic training. In particular, what factors helped contribute to their rise to prominence? (Malasarn, Bloom, & Crumpton, 2002, p. 56)

Determining the Study's Significance

We have attended many meetings in which the first question to a researcher was, "So what? Why is this study important, or worth doing?" The question of significance emerges every time students or researchers raise an idea for a study. Be sure to plan for this question as you conceptualize your study.

Studies are significant for many reasons. As an example, the findings of a study may improve children's health habits. Perhaps a study's outcome could improve a school district's funding for physical education. Maybe the research results highlight the voice of population that is often ignored. Marshall & Rossman (1999) state that studies are significant when they affect practice, policy, or theory.

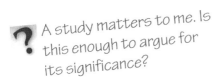
A study matters to me. Is this enough to argue for its significance?

This is a good beginning, but you must stretch yourself a little bit more. Ask yourself why the study is significant to you. Once you answer that question, you may be able to place your study in a broader context.

Assembling a Literature Review

We mentioned earlier that researchers often draft an introduction and conduct a literature review simultaneously. It is very difficult to articulate a conceptual framework and identify a researchable problem without reviewing related literature. Therefore, researchers often revise drafts of the introduction, problem statement, and purpose statement as their understanding of the supporting literature deepens. This section explains what constitutes a literature review, and then clarifies which sources of literature are most appropriate for conceptualizing your study.

Reviewing Literature

A literature review synthesizes, compares, and contrasts data from published sources. It is also essential for constructing your conceptual framework.

Your review of the literature demonstrates your grasp of a particular topic and your familiarity with related research. You must identify what is already known on a subject to determine whether your study is original and significant. As you begin the process, decide what type of literature to review. You should also think about how much literature you will review and how you plan to find it.

Navigating the Process

Your first decision as you conceptualize your study is what literature to review and include. Because the amount of information available in print is expanding at a tremendous rate, you can probably find published information on most topics. However, research studies must be informed by reputable and reliable sources.

 When building my case, should I include a full literature review or focus on the key articles?

Initially, you should include a full literature review. You may focus on key pieces when you develop an article from your study. See chapter 8 for more information.

Reputable and reliable sources include peer-reviewed (or refereed) articles from scholarly journals, peer-reviewed abstracts that have been published, texts written by experts or scholars in a particular discipline, and peer-reviewed papers from professional conferences. Peer-reviewed articles or abstracts are those that have been scrutinized by professionals from the field before publication. The review process screens out manuscripts that do not meet a professional community's standards of quality for publication. Many search indexes identify whether a journal article is peer-reviewed, but if you are uncertain, check to see if the journal has an editorial board. If the journal lists names and credentials of editorial-board members, it likely contains refereed articles.

Another important consideration when reviewing literature is whether sources are primary or secondary. In primary sources, authors present an original idea or collect and analyze the data for the study themselves. Research-based journal articles are an example. In secondary sources, authors examine primary sources and then summarize them, explain them, or compare them with other studies. If possible, go to the primary source and examine the information yourself. The danger of using secondary sources is that someone else's interpretation of primary information could be skewed, altered, or misinterpreted.

The final consideration is the age of the literature you review. Are you looking for literature published in the last 5, 10, or 15 years? The answer to this tricky question depends on the nature of your topic. If your conceptual framework is based on traditional theory such as Bandura's self-efficacy theory, you should review Bandura's original work on the topic from 1977. However, stick with literature that is more recent for contemporary issues. Go back as far as necessary to find literature that supports a compelling argument, but keep in mind that many readers want some assurance that you are aware of the latest professional developments. You should include some literature from this millennium in your review.

Informing Your Research Problem With the Literature Review

Your literature review is intertwined with your study, and will change throughout the writing process as you reexamine your sources and revise your statements. The process is cyclical: some ideas stimulate your search of the literature and some findings from the literature inform your study. Subsequently, your thinking evolves over time. It will take time, but the process is an enriching and important part of your research journey.

 I already know how I want to collect my data and whom my research participants will be. Can I skip the process of building a conceptual framework and writing research questions and go straight to the methods section?

No! Beginning a study by collecting data is dangerous because your research will be driven by the methods and convenience of the sample rather than a theoretical base. Remember, research questions should drive research!

Summary

In order to conceptualize your study, you must write a draft of your introduction that is informed by your literature review. It should include background information, a problem statement, purpose statement, research questions, and the significance of the study. The introduction will guide your process of data collection and analysis.

CONTINUING YOUR EDUCATIONAL JOURNEY

Learning Through Activity

1. Read the introduction to "Darwinism in the Gym" from appendix B, and identify the theoretical framework and the two goal (purpose) statements.

2. After identifying the goal statements in the question above, write two or three research questions for this study.

3. Locate a qualitative study that is not included in this text and list its purpose, conceptual framework, and significance.

 Checking Your Knowledge

1. A conceptual framework is based on
   a. existing theory
   b. the significance of the study
   c. a discipline's literature base
   d. a and b
   e. a and c

2. Research questions involve a list of questions you will ask participants in an interview.
   a. true
   b. false

3. Which of the following is the best form of evidence to use in a literature review?
   a. a peer-reviewed, research-based article from a scholarly journal
   b. a textbook
   c. a journal article that has not been reviewed by peers
   d. objective evidence
   e. a and c

4. Which of the following items found in scholarly literature is a primary source of information?
   a. an article reviewing the research literature
   b. an encyclopedia
   c. a research-based article
   d. a textbook

5. Which of the following items is not a component of an introduction?
   a. significance of the study
   b. methods
   c. problem statement
   d. purpose statement
   e. research questions

 Thinking About It

1. You are interested in studying how coaches cope with frequent moves and job changes. What key terms would you use to search for pertinent literature?

2. Identify two or three reasons why a researcher would study the experiences of female athletic directors in NCAA Division I programs.

3. Research questions guide a study and help narrow a researcher's focus. A researcher may use one broad research question or several specific research questions. List the advantages and disadvantages for each approach.

 Making a Stretch

These readings will expand your knowledge on initially planning and introducing a study.

Booth, W.C., Colomb, G.G., & Williams, J.M. (2003). *The craft of research* (2nd ed.). Chicago: University of Chicago.

Marshall, C., & Rossman, G.B. (1999). *Designing qualitative research* (3rd ed.). Thousand Oaks, CA: Sage.

Collecting and Analyzing Qualitative Data

Learning Objectives

Readers will be able to do the following:

1. Identify various forms of qualitative data.
2. Collect qualitative data through interviews, observations, and document reviews.
3. Outline different sampling strategies.
4. Explain the steps of qualitative data analysis.

Explaining the Interpretive Process

Many readers who are unfamiliar with qualitative inquiry find its method of data collection and analysis ghostly or magical, as if researchers simply pull a rabbit out of a hat. They may think the process lacks structure, discipline, and thought. As you learned in chapter 3, however, the procedures for collecting and analyzing data require a great deal of planning. This chapter explains different methods of data collection and outlines the steps of data analysis. We begin with an overview of the interpretive process.

Interpretation, or the process of explaining meaning, is the underlying premise to all qualitative research, which seeks understanding about particular problems. Stringer (2004) explains that the process of interpretation allows researchers to understand the experiences of participants, specifically what they feel and believe about various phenomena. Researchers gain insight from the experiences of their participants by systematically collecting and analyzing appropriate data. The outcome of any inquiry is only as good as the data you collect. The first step in obtaining quality information is to select an appropriate sample of participants.

Sampling and Participants

Once you have developed your study's purpose, research questions, and significance, you are ready to begin the process of participant sampling. You must identify whom to choose, how to choose, and how many to choose. Qualitative research relies on strategies of purposeful sampling (Patton, 1990; Creswell, 2005), or the selection of participants who can provide data related to the study's purpose and research questions. Tuckett (2004) explains that the process of participant sampling should be fully examined for every qualitative study.

Remember that you will make sampling decisions both when you create your proposal and as the study progresses. Qualitative research is called emergent because some findings prompt you to look for confirming data or explore aspects of a phenomenon that you did not anticipate. In these cases, you may need to interview participants whom you did not originally consider.

When choosing your sampling strategy, you must first reflect on your study's purpose and research questions. Imagine that the purpose of your study is to "explore the physical-education experiences of early-adolescent students who are obese." Because you are interested in a very specific population, you will select participants purposefully, likely recruiting adolescent students who are both obese and enrolled in a physical education class. Your study proposal should define the terms *early adolescence* and *obesity* so you can justify the inclusion of these participants in your sample. You may later consider including other participants to enrich the context of the study. For example, data from physical education teachers may help you better understand the classroom setting.

Types of Sampling

Many subtypes exist within the general category of purposeful sampling (Creswell, 2007; Patton, 1990; Higginbottom, 2004; Coyne, 1997; Morse, 1991). Here are several examples:

- *Criterion sampling.* In this type of sampling, also called criterion-based selection (Maxwell, 1996), the researcher predetermines a set of criteria for selecting participants. In other words, participants are chosen because they have a particular feature, attribute, or characteristic, or have had a specific experience. In the preceding example, criterion-based selection would help you recruit the following participants:

  1. Students between the ages of 12 and 14.
  2. Students currently enrolled in a physical education class.
  3. Females who have higher than 30% body fat or males who have higher than 25% body fat.

- *Typical sampling.* In typical sampling, a researcher or research team chooses participants who fit with a norm for a given population. As Creswell (1998) notes, it can be difficult for researchers to identify what is typical for a certain group. According to Merriam, the typical sample should be "selected because it reflects the average person, situation, or instance of the phenomenon of interest" (1998, p. 62). Researchers who use typical sampling should choose participants who are average (Patton, 1990).

- *Maximum variation sampling.* In maximum variation sampling, researchers explore different perspectives of one situation by recruiting people from a variety of backgrounds for a study (Byrne, 2001). This form of sampling helps researchers fully explore many facets of a problem and investigate issues holistically.

Several considerations exist for maximum variation sampling. Refer back to the example of the previous study, in which the purpose is to "explore the physical-education experiences of early-adolescent students who are obese." Researchers may first wish to maximize the variation by including participants of both genders. Next, recognizing that other cultures view obesity differently, they may include participants from different cultures or races. Finally, because socioeconomic status may be an influential factor in obesity, they may include participants from lower-, middle-, and upper-class backgrounds.

- *Deviant sampling.* In deviant sampling, researchers intentionally attempt to find participants, programs, or situations that deviate substantially from the norm. In other words, researchers examine situations that are unusual or distinctly nonaverage. Imagine that an elementary school in a particularly urban environment implements a comprehensive wellness program at a time when most schools place their limited resources in other departments. A researcher could frame this program as atypical and conduct a study to learn about the program's purpose, structure, and function so that other schools might choose a similar program.

- *Snowball sampling.* Snowball sampling is called *chain, network* (Merriam, 1998), or *nominated sampling.* In this form of sampling, researchers rely on participants to direct them toward others who meet the study's criteria. Imagine that you are planning to examine why physical educators drop out of their roles. For this sample, you wish to find teachers who worked at a high school for 1-5 years before quitting. When you propose your study, you think you will need six or seven participants, but you know only four people who fit the criteria. While interviewing the four original participants, you ask them if they know any other former physical educators who stopped teaching after one or more years. In this example, you are relying on snowball sampling, hoping that one participant will lead you to another. One limitation of this approach is that all of a study's participants may be somehow connected with each other. If possible, start the process of snowball sampling with at least two different participants.

- *Total population sampling.* Total population sampling includes everyone associated with a small group as participants in a study. For example, if you were conducting a case study of a unique doctoral degree being offered to a small group of students in the field of health education, you might cover the total population by interviewing the teacher, the students, and the department chair.

Variations in Sampling Strategies

The direction of your study and the exploratory nature of your method will likely lead you to mix your sampling strategies. Consider a study by Pitney (2008) designed to explore the perceptions of professional commitment among athletic trainers who work in high schools. Pitney needed to interview athletic trainers who had worked in this setting for some time and who personally believed they were committed to their professional roles. Using the membership services of the National Athletic Trainers' Association (NATA), Pitney purchased a random sample of 1000 members who fit specific criteria:

1. They had worked in a high school for at least 10 years.
2. They self-identified as being committed to their professional roles.

After obtaining the names of members from the NATA, Pitney sent an initial e-mail inviting them to volunteer for the study. He then chose participants to interview from the group of respondents. In this real-life example, Pitney initially combined random sampling with criterion sampling, and then identified individual participants. Random sampling is more common in quantitative research, but it can effectively identify an initial group of potential participants for a qualitative study. Many additional sampling strategies exist. Refer to Patton (1990) or Creswell (1998 & 2007) for more information.

Sample Size

Although qualitative studies generally use a small number of participants, sample size is still important. You will always be asked about your sample size when you share your proposal with others. Even if you don't have a specific answer to the question, "How many participants will you include in your study?" you can respond appropriately by saying, "Enough to address the study's purpose and to answer the research questions." You could also say, "Enough to gain insight about the phenomenon under investigation."

The general principle that guides sample size is saturation of data, or redundancy of data. This is the point at which you no longer encounter new information, or continually encounter the same information. You will know this moment when it happens. Imagine you are enrolled in three different research courses. In each course, you learn that the study's purpose and questions guide your methods and design. After hearing this concept in three different classes, you may consider it redundant. You have reached your saturation point. When you begin to hear similar information from multiple participants in a study, your qualitative research may have reached its saturation point.

When writing a research proposal, you do not know when saturation will occur. Therefore, you should overestimate the number of participants for your study (Morse, 2000). Ultimately, an inverse relationship exists between the quality of the data received from participants and the number of participants needed for a study (Morse, 2000). The higher the quality of usable data obtained from each participant, the fewer interviews you will need to conduct.

In addition to the general principle of data saturation, you may find it helpful to examine published qualitative studies to get a sense of participant numbers. In the following examples, the sample sizes range from 7 to 36 participants:

- Podlog & Eklund (2006) investigated 12 competitive athletes' return to sports after serious injuries.
- Young, White, & McTeer (1994) examined the subjective experiences of sport injuries among 16 male athletes.
- Chiang (2005) examined the quality of life of children with asthma by interviewing 11 children and their parents.
- Mensch & Ennis (2002) explored the strategies thought to enhance learning in athletic-training education programs by interviewing 21 students and 12 instructors, for a total of 33 participants.
- Graham, Dugdill, & Cable (2005) studied the perspectives of 12 health professionals (10 general practitioners and 2 nurse practitioners) on the exercise-referral process.
- Lempp & Seale (2004) investigated 36 medical students' perceptions of hidden curriculums.

- Malasarn, Bloom, & Crumpton (2002) examined the development of 7 men who were NCAA-Division-I-certified athletic trainers.

Sample Accessibility

Another important consideration in qualitative research is how you will gain access to individuals or programs to collect your data. When selecting a research site and participants, you are often at the mercy of factors beyond your control. This may affect your sampling strategy and sample size. Imagine that you have planned your study and identified potential participants, but are denied access to your chosen group of people. You will need to be creative and resourceful in determining where and how to recruit alternative participants. Take advantage of the flexible and emergent nature of qualitative research. You may need to change your sampling strategy and sample size to complete the study.

Collecting Qualitative Data

Qualitative researchers use many forms of data. Those who are interested in learning about specific cultures might consider artifacts and ritualistic items. However, most of us use interviews, observations, and documents.

Conducting Effective Interviews

Interviews are perhaps the most common source of qualitative data. They let researchers enter the world of their participants and learn rich and valuable information about their experiences. What better method exists for understanding someone's beliefs, attitudes, and perceptions?

However, in order to fully understand your participants' views, you must conduct your interviews effectively. Effective interviews address the central aspects of the study and attempt to collect information related to all research questions. Interviews must also make your participants feel comfortable and open to sharing their views. Many authors have offered thoughts about how interviews should be conducted. We have formulated the following section based on the works of Schatsman and Strauss (1973) and Kvale (1996). We use their advice to guide our own interviews.

Organizing Your Interview

Remember that you are about to collect critical information that will help you answer a research question. Moreover, you have been invited to share a small portion of your participants' world, and they have offered you their valuable time. You must plan ahead to make every moment with them count and ensure that the event is mutually enriching.

When planning for your interview, consider these five important components:

1. *Setting the stage*. Provide the interviewee with a framework for the study.
2. *Building the relationship*. Set the tone for your relationship and help your interviewee feel comfortable.
3. *Addressing the research focus*. Ask your interview questions, and then follow up with probes to obtain a depth of information.

4. *Debriefing.* Conclude the interview by identifying the important things you learned. Obtain permission to follow up at another time if necessary.

5. *Thanking your participants.* Genuinely thank your participants for the opportunity to learn about their experiences.

Setting the Stage   You have an obligation to orient your participants to the interview process and share why you are interested in speaking with them. Take time to briefly explain the purpose of your study in terms they will understand and remind them that their participation is voluntary. Ask them for permission to record the conversation and let them know that they can decline to answer a question at any time. Reassure your participants that you will maintain confidentiality before you begin the interview, and then ask them to sign an informed consent form. Chapter 6 discusses these ethical issues in more detail.

Building the Relationship   You should begin the interview by setting the tone for your relationship with the participants. Help them feel comfortable with you by easing them into the conversation. Kvale (1996) suggests using a series of dynamic questions to establish a positive relationship with participants. Some examples of dynamic questions include, "Tell me a little bit about yourself," "What is your academic background?" or even "Can you start by telling me about a typical day for you?" These relational questions should be easy to answer and should give participants an opportunity to talk about themselves that does not stray too far from the research topic. This portion of the interview is also a good time to discover some demographic questions, such as their amount of job experience or the highest degree they earned.

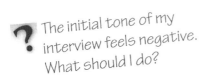
The initial tone of my interview feels negative. What should I do?

Look for the origin of the negative feelings. Is the tone coming from you, the participant, or external factors? Next, address the issue. Stop the interview and explore the participant's experience. It is possible that for whatever reason, you are an intrusion at that moment in time. Although the participant may be unable or unwilling to articulate this, you must determine whether the interview should continue as planned. Reschedule the interview if necessary.

Addressing the Research Focus   The research focus is the heart of the process, your reason for interviewing the participant in the first place. Your goal is to learn information that will help you address your research questions. In an interview sandwich, this section would surely be the meat. You must ask good questions, probe appropriately to obtain a depth of information, and direct participants to share information.

Construct your interview questions or statements according to the purpose of your study. Kvale (1996) calls them *thematic questions* because they relate specifically to the research theme. When you draft your interview questions, begin by reading through your research questions. Your interview question should differ from your research questions as they address the research theme or focus at a different level of abstraction. As an example, suppose that your central research question is, "What organizational factors influence the professional commitment of physical education teachers?" You may need to ask many specific questions about professional commitment during interviews in order to gain a full understanding of this phenomenon in context. If you ask teachers during interviews what organizational factors influence their ability to stay committed to teaching, you might get responses that are rather short and superficial. But if you develop a battery of questions, you are more likely to gain true insight about a phenomenon. Consider the following series of questions or statements:

1. Describe for me what it means to be committed to your role as a teacher.
2. How have you maintained your commitment to teaching?
3. Describe for me a situation that challenged your commitment to teaching. How did you deal with that situation?
4. How does your work environment specifically influence your commitment?
5. How have some of your colleagues in physical education maintained their commitment?
6. What advice would you give to new physical education teachers to help them maintain their enthusiasm for the job over the course of their careers?

As you can see, these questions explore the same phenomenon in slightly different ways, allowing the researcher to more fully develop an understanding of the participants' perceptions. When planning a study, pay attention to how you address the research purpose and questions in the interview. Remember that interview questions should serve one of two purposes: to develop a relationship with the participant, or to understand their perceptions as they relate to the purpose of your study.

You may also need to elicit information from the interviewee by asking them to respond to statements. Consider the following examples:

• Describe for me how you deal with noncompliant athletes.
• Explain how you work with athletes whom you label as noncompliant.

Both examples address the same issue in an open-ended manner, and either can be used effectively in an interview. It is a good idea to integrate both techniques into your interview.

What good is an interview if you ask bad questions or fail to guide participants to share their perspectives? If you ask poor questions or fail to garner information related to the research topic, you can ruin your study. Remember that the collection of bad data leads to the analysis of bad data, which can result in questionable research findings. With this in mind, plan out your interviews carefully, paying particular attention to the questions you will ask. Many researchers prepare an interview guide as part of this process.

Interview guides, or interview protocols (Devers & Frankel, 2000), list questions and statements for researchers to follow. Three types of guides exist: unstructured, semistructured, and structured. An unstructured guide may provide an opening question or statement to ask participants, but the researcher is responsible for creating the majority of interview questions on the spot. Novice researchers may view this approach as risky. It can be difficult to create meaningful questions during an interview, especially if your participant does not seem open to sharing information. However, the unstructured guide allows experienced researchers a great deal of freedom to explore a phenomenon.

The structured interview guide is the easiest to follow, but is also the most confining. It contains a list of questions to be asked of all participants. Researchers may not ask additional questions or probes during the interview to further explore a topic.

The semistructured interview guide, which we recommend most highly for qualitative research, contains a series of standard questions to ask all participants. However, this guide also lets researchers probe participants and add questions as needed. It provides enough structure to walk a novice researcher through an interview, but also allows for tangential discussions about critical and noteworthy information. The following is an example of a semistructured interview guide from a study on role strain by Pitney, Stuart, and Parker (2008).

Semistructured Interview Guide

- Describe a typical day as a teacher and athletic trainer.
- Describe a typical week as a teacher and trainer.
- What drew you to the dual-role position?
- What support do you receive in your dual role?
- What are your responsibilities as an athletic trainer?
- How would you describe your teaching responsibilities?
- How do you feel your dual role is viewed by _____?
    - probes
        - administrators
        - colleagues
        - athletes
        - students

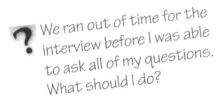 We ran out of time for the interview before I was able to ask all of my questions. What should I do?

The duration of interviews varies because some participants talk a lot and others provide brief answers. If you run out of time, ask to schedule another interview.

As you develop your interview guide, be sure to include different types of questions. Just as readers appreciate variations of sentence structure, participants in an interview may need questions to be asked in a variety of ways. Suppose you are interviewing physicians about their perceptions of obesity and begin each interview with the following three questions:

- What does the term obesity mean to you?
- What does it mean to you to have many patients who are obese?
- What does being obese mean for your patients?

The similar structure of each question may seem repetitive to your participants, which may affect the depth of their responses.

Include both open- and closed-ended questions in your interview guide. As the names suggest, open-ended questions encourage descriptive responses and closed-ended questions can usually be answered with yes or no. Some topics may require closed-ended questions, but be sure to follow them up with a probe such as, "Can you explain why?" Consider the following interview question from a study by Podlog and Eklund (2006) on competitive athletes' return to sports after injuries. Although they ask a closed-ended question, they immediately follow it up with a clarifying question: "Have you had any physical, mental, or technical setbacks/adversity since returning [to your sport]? If so, what have you done to deal with these fears/concerns?" (p. 48).

 The interview is "off course." How do I re-direct my participant without cutting her off and feeling as though her voice is not being heard?

Although you should be wary of interrupting or ignoring your participant's train of thought, it may help to use a phrase like, "Can you relate that thought to the initial question for me?" You could also say something like, "I would really like to hear more about this topic. Would you mind if we revisit it at the end of the interview?"

Debriefing   This portion brings closure to the interview. It is your chance to share key points that you have learned and clarify your thoughts about the participants' perceptions. At this time, obtain permission to follow up with another interview if necessary. Be sure to let your participants know that you will likely use some of their data in a research report. However, if you use their information, you will identify them with pseudonyms (fake name) rather than their real names. Participants may select their own pseudonyms at this time if they prefer.

 My participant asked me to keep part of the interview off the record before revealing information that is critical to my study. Can I use the information?

The short answer is no! However, you may ask participants why they wish to exclude certain comments from the interview and offer them reassurances about your confidentiality practices. If your participants remain firm in their convictions, you must honor their requests.

Thanking Your Participants   This is the final phase of your interview. Your participants have shared their time and experiences to assist you with your research questions. Kindly express your gratitude for being allowed into their personal lives.

Conducting Focus Groups

Depending on the situation, you may want to collect data by conducting a group interview, or focus group. Focus groups let you obtain a lot of divergent thoughts over a brief period of time. This can be helpful when exploring participants' insights. For example, in a study examining perceptions of obesity prevention, healthy eating, and activity, Hesketh, Waters, Green, Salmon, and Williams (2005) conducted focus groups at three different schools. Focus groups were the researchers' sole method of data collection, and they conducted different groups for parents and children.

Researchers may prefer different sizes of groups depending on the situation. Stringer (2004) states that focus groups work best with four to six participants, but Loeb, Penrod, & Hupcey (2006) identified six to eight participants as ideal for their studies. Groups with 20 or more participants are hard to manage. Researchers are rarely able to involve everyone in the discussion. Participants may lose interest while waiting for so many others to share their thoughts (Loeb et al., 2006). Also, some people may be less willing to share their experiences in large groups. On the other hand, groups that are too small may not yield many divergent perspectives.

The advantage of a focus group discussion is that it can stimulate the thoughts of group members. After one participant responds to a question, another may agree, disagree, or provide a new insight. The disadvantage of focus groups is that one person can monopolize the conversation if the group is managed poorly.

The general guidelines for one-on-one interviews also work for focus groups. However, you should broaden the section on building relationships to establish good group dynamics. Be sure to learn something about everyone in the group. Consider distributing name tags so that participants can call one another by name. You should establish ground rules about respect and confidentiality (Loeb et al., 2006). During the discussion, make a conscious effort to involve everyone and maintain the flow of conversation. Finally, be sure to effectively record your findings.

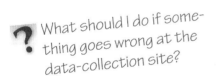
What should I do if something goes wrong at the data-collection site?

The general rule of thumb is to plan for every potential problem. Pack extra batteries for recording devices and both hard and electronic copies of your interview protocol. Allow more time than you need to travel to the site, and be pleasantly surprised if you don't get caught in traffic! If your interviewee does not meet you, double check your schedule and follow up with a very polite phone call or e-mail. Remember you are the guest, so smile!

Observing Participants

Observations are another common form of data collection in qualitative research. Although it is helpful to learn about peoples' perceptions during the course of an interview, observing their behaviors adds a whole new dimension to a study. Suppose you are conducting a study to explore coaches' use of reinforcement. During an interview, one coach says that she primarily uses positive reinforcement with her athletes and uses negative reinforcement very rarely. When you observe her in practice, however, she continually uses demeaning phrases. In this example, your observations illuminated information that is inconsistent with what you learned from the interview. You may need to conduct an additional interview to clarify what you have learned. Sometimes, you must consider many forms of data to answer your research questions.

Being an Effective Observer

Observations allow you to immerse yourself in a participant's natural setting to learn about a specific context (Pitney & Parker, 2001). As an effective observer you must address the following three components before initiating observations:

1. Decide how the observations will be documented.
2. Determine what, exactly, will be observed.
3. Establish the role you will take as observer.

Documenting Your Observations

In an extensive study, you probably will not remember all that you see. Therefore, you must keep good records, including documentation of your observations. Remember that each researcher prefers to handle information obtained from an observation differently. Some researchers create a general narrative about the observation experience, while others might like to explain the events minute by minute. You should determine the method that works best for you.

Let us return to the example of the study of positive and negative feedback used by coaches in athletic practices. You could create an observational instrument that categorizes interactions between coaches and athletes as either positive or negative, and leaves room for comments about speech and behavior. At the end of your observation, you would have a better sense of which type of feedback is dominant during practice.

Remember that data collection and analysis is an interpretive process. As you collect data through observations, you will begin to formulate ideas about its meaning. In other words, as you observe particular interactions, you will start to relate them back to your study's purpose and research questions. Therefore, you must record not only what you see, but also any thoughts you have about what your observations mean. Note your interpretations twice: once when the behavior occurs, and again after observation. Be cautious to first record what actually took place before adding your interpretations. At times, you may be sorely tempted to go straight to the interpretation—resist, resist, resist! If you begin writing your interpretations during an observation, you will likely miss the next event that occurs.

Knowing What to Observe

Your research questions and the purpose of your study determine your focus during observations. Qualitative researchers generally choose to observe the physical setting, including the participants' behaviors, actions, interactions with others, and speech. When observing a practice session to explore a coach's use of reinforcement, you would pay particular attention to what the coach says. You might also note her body language, such as rolling her eyes, shaking her head, or waving her hands. Document any behaviors that could be significant in the context of reinforcement.

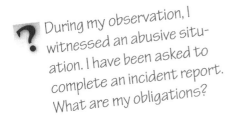

During my observation, I witnessed an abusive situation. I have been asked to complete an incident report. What are my obligations?

Chapter 6 discusses this scenario in more detail. Remember that your primary concern is the safety of your participants. Follow the policies in place, regardless of whether it may jeopardize your access to data.

Knowing How to Observe

You will need to choose how to perform your observations. Merriam (1998) identifies that observations can be viewed on a continuum that ranges from participatory to non-participatory (see figure 4.1). Similarly, Mills (2007) explains the role of the observer as ranging from active observation to passive observation. Each form of observation has advantages and disadvantages.

If you choose to observe passively, you will have time to document events as they occur. A disadvantage of assuming a passive role is that your presence may be considered intrusive. For example, athletes and athletic trainers may be open to your presence in the athletic training room as long as you are involved in various activities. Simply sitting yourself in the corner of the room with a clipboard makes you conspicuous, and the participants may alter their behavior.

If you choose to observe as an active or complete participant, you will have the advantage of fitting in and becoming one of the players in your chosen context. In the previous example of the athletic training room, as a complete participant you might assist with taping, bandaging, or caring for wounds. Your presence will be less conspicuous, thus helping participants feel more comfortable and free to be themselves. A disadvantage of this approach is that you must remember the events you observe and document them later. You are limited by the strength of your memory. Your involvement may also be limited by your expertise; don't pretend to know how to tape an ankle just so you can be involved in the setting!

Some researchers strike a good balance with a mixture of both observational roles. In some contexts, you can be part participant and part observer. This approach helps you fit in, but still allows you to observe and document your findings when appropriate.

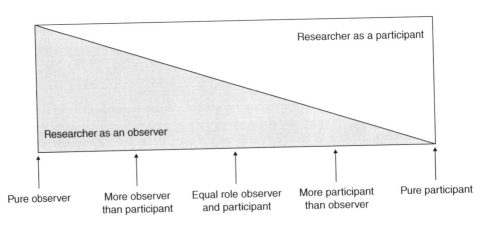

Figure 4.1 The observation continuum.

Regardless of your method, remember that you are observing a natural setting; your goal is to avoid disrupting it in any way. You must be tactful to fit into the setting, and you must be intuitive to discern when your presence becomes intrusive.

The final consideration for observations is whether to record the events using digital media or videotape. This practice is advantageous because it provides a permanent record of events that you can review later. The recorder can also act as an extension of your senses, allowing you to capture more details than you might otherwise see and hear. This practice also has its disadvantages. Cameras can be seen as intrusive. Participants may feel reluctant to be recorded or change their behavior for the video. It may also be inappropriate to record in certain settings, such as an athletic training room. Ask your contact person at the research site whether this type of recording would be possible.

Using Documents as Data

Qualitative researchers also commonly use documents as data. Examples of documents that researchers may collect in the professions of health and physical activity include medical records, corporate memos, lesson plans, syllabi, court documents, and letters. Although rarely used as primary data, documents often contain information that helps researchers clarify findings from interviews and observations. For example, we are currently conducting a study that examines preprofessional students' (students of athletic training and physical education) perceptions of obesity, as well as how students are instructed to interact with people who are obese. We supplemented our interviews by examining the syllabi of physical education classes to identify the course objectives and content. We believe this combined approach will yield substantial information and provide us with a comprehensive understanding of the issue.

Obtaining Alternative Data

Alternative forms of qualitative data include photographs, videotapes, and artifacts. Suppose you are conducting a qualitative study on the leadership styles of highly successful professional coaches. After interviewing the coaches about their leadership beliefs and values, you might also observe their practices. If a coach offers to show you video tapes of pregame and halftime talks from the last season, it is simply the icing on the cake. Such evidence could be very informative when combined with the other data.

Ideally, you will collect data in a variety of ways that inform your interpretation and address your research questions. It would be awesome to have mountains of data at your disposal, but this is rarely the case. Many factors affect your method of data collection, as well as the type and amount of data you choose to collect. Considerations include your time frame, available monetary resources, and access to various contexts and people. You must also consider the strengths and weaknesses of each approach and how each addresses your purpose and research questions.

Once you begin to collect data, you also begin the process of analysis. Remember that you will intuitively begin to analyze even the smallest amount of qualitative data. In fact, the process of data collection and analysis is continuous and ongoing (Pitney & Parker, 2001). Qualitative research allows you to analyze your data as you collect it. This differs from quantitative research, in which you first collect all of your data, and then analyze it at the end. Over time, you will have a great deal of data to analyze, but the process of analysis is fascinating!

Analyzing Qualitative Data

Qualitative data analysis is an interpretive event. In other words, researchers collect and analyze data concurrently, and then infer meaning and draw reasonable insights to answer the research questions.

Let us discuss in more detail the process of collecting and analyzing data concurrently. Remember that qualitative researchers attempt to understand the meaning people assign to their experiences. They require an instrument for analysis that is sensitive to speech, writing, and human behaviors to draw conclusions about the qualitative dimension of participants' lives. Therefore, the researcher serves as the instrument for data analysis. From the moment participants share their thoughts, experiences, perspectives, and perceptions, researchers cannot help but wonder what they mean and how they relate to the experience of others.

The process of data analysis requires researchers to use inductive reasoning to categorize information. The process is both creative and structured. We present eight steps of qualitative analysis based on our research experience and qualitative literature. The process follows the acronym CREATIVE:

The Eight CREATIVE Steps of Data Analysis

Consider the study's research questions and purpose statement.

Read through your transcripts to gain a holistic sense of the data.

Examine the data for information related to your research questions.

Assign labels to these units of information that capture their meaning.

Thematize the data.

Interpret the emergent themes as they relate to the study's research questions and purpose statement.

Verify the trustworthiness of your findings.

Engage in the writing process to describe your findings.

Consider Your Research Questions and Purpose

Once you have collected some data, you are ready to start the process of systematic analysis. You can begin analysis before you finish the process of data collection. First, consider your study's purpose and research questions. Although this step may sound odd, it will help you focus on critical information. You will later look for data patterns among interviews or observations. When examining the data, you may feel as though all the information is critical and noteworthy. Therefore, you must use the study's purpose as a lens when reading transcripts, documents, and observation notes.

Read Through Your Transcripts

Once you have revisited your research questions and purpose statement, you are ready to read your data. Read through your data once without judging the content. Creswell (2005) calls this step *preliminary exploratory analysis.* He states that the process of exploration helps researchers gain a general sense of the data. As you read, make mental or literal notes to yourself about the significance of the content, but try to reserve your judgment until you have examined all of the data.

Examine Data for Important Information

As you read through your documents to gain a sense of the data, you will probably encounter information of significance to the study. At this point, you may highlight, bracket, underline, or otherwise tag information that addresses your research questions.

Meaningful information can come in a variety of forms, including a single word, a single sentence, multiple sentences, a paragraph of text, several paragraphs, or multiple pages of text. Work to identify information that will help you answer your research questions. Stringer (2004) calls this step *unitizing the data*. He discusses the unitizing process and states that the units of data become the building blocks of the research findings. The researcher's job is to logically arrange the blocks to interpret the context.

Various forms of qualitative research use different terms for the information identified from the data. Some call the information *concepts* (Strauss & Corbin, 1990), while others use *meaning units* or *units of data* (Lincoln & Guba, 1985). This part of analysis is often performed simultaneously with the next step, assigning labels.

Assign Labels

After identifying units of information, you have a large volume of text to organize. The words, sentences, and paragraphs you find meaningful will accumulate quickly, so you must have a strategy for managing your data. Label the units of information with a term or code that captures its meaning. This labeling process is often called *coding* (Creswell, 2005).

For example, consider a qualitative study in which the purpose is to "gain insight and understanding related to how professionals learn their organizational role." Figure 4.2 provides an excerpt from a study's transcripts in which important information is identified (step 3 of the analysis process) and given labels that signify its meaning (step 4 in the analysis process).

 The transcription of my data is 500 pages long. Help!

Welcome to the pros and cons of data collection. The upside is you have lots of good data to choose from, and the downside is that you must work with 500 pages of data! If you read your transcripts as you collect the data, you will be familiar with most of the pages. Remember, you will not use all of the data. Read your transcripts through to get a sense of the whole and direct you to a starting point, which is often not at the beginning!

Thematize the Data

Once you have assigned labels or codes, you are ready to further analyze your findings. One of the universal steps of qualitative research is examining data for patterns. This step is called *thematization*. However, you will probably begin identifying patterns intuitively as you assign labels.

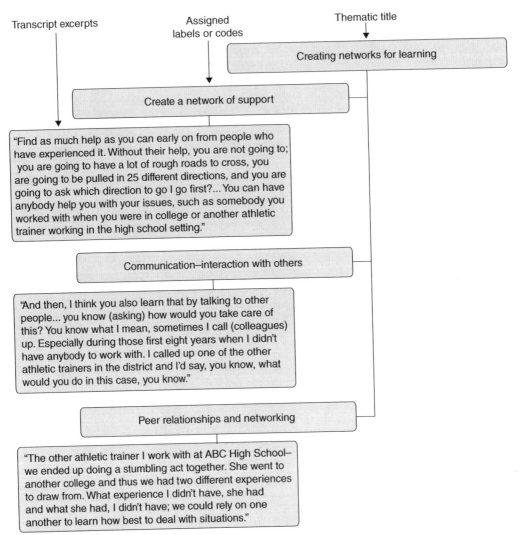

Figure 4.2 This is an analysis of a transcript from an actual study. The excerpts are samples of data identified as important in terms of the study's purpose (Pitney, 2002). The quotes on the left are those containing important information related to the study's purpose; the middle section displays the assigned labels. The codes are placed into themes based on similarities.

In the thematization step, researchers organize data into categories, or themes. In other words, you will pick through your labels and group those that are similar or hang together. The terms *categories* and *themes* are often used interchangeably, but this can be confusing for readers. We recommend choosing one term and using it consistently.

A critical aspect of thematizing your data is to consider if and how the emerging themes are related. In many cases, initial themes may be closely related to other themes. Look into any developing relationships that you notice. You may need to combine related themes to clearly capture the meaning of the data. For example, in a study examining the mentoring process of athletic-training students, Pitney and Ehlers (2004) initially identified nine emergent categories (themes). While examining the relationship between the categories, they found that three categories were very closely related to the beginning of the mentoring connection: "mentor approachability," "mentor accessibility," and "protégé initiative." The authors grouped these three categories under the heading, "mentoring prerequisites," and presented the original groups as subcategories.

 While examining my data, I realized that one piece could be included in two different themes. How should I proceed?

If this happens often and more than one piece of data recurs in the two themes, consider combining them into one. You can also place the same piece of data in both themes as long as you clearly justify your decision.

Interpret Themes

The themes that emerge from your data are your primary findings. You should examine them and give them titles that capture their meaning. The title of the theme should relate to the purpose of your study and have relevance for the readers. Moreover, the thematic titles and supporting data should describe the phenomenon or explain the process under investigation. We again use the term *interpret* because our mission with qualitative research is to gain understanding and infer meaning from the data. If your themes do not identify definitive patterns, then you have not completed your mission. You must continue to collect and analyze data.

Data should be collected until you either notice redundant findings or stop discovering new information. We referred to this process earlier as *saturation of data*. When you notice this pattern and can fully answer your research questions, you may conclude the process of data collection and analysis.

Verify your Findings

An important aspect of the analysis process is conducting verification procedures (Miles & Huberman, 1984). Although we present these strategies in more detail later, we mention them here because they have a substantial bearing on your analysis.

As you gain insight and understanding about the phenomenon you are studying and draw conclusions related to your research questions, you must be confident that your findings are accurate. You can gain confidence in a variety of ways, including member checks, peer debriefing, and triangulation. Chapter 5 thoroughly explains these strategies. For now, note that verifying your findings can significantly influence the strength of your analysis.

Engage in the Writing Process

You must spend time thinking about your findings. Perhaps the best way to stimulate deep and meaningful thinking is to engage in the writing process. Whether you are writing observation notes, research notes to yourself, or explanations of your findings, the act of writing forces you to articulate what you know or have learned about a phenomenon. For this reason, we include writing as a critical step in the analysis process.

The writing process helps you reexamine your themes and the relationships among them. It also highlights any need to change titles or combine themes in order to better represent the data. In the early stages, your writing is tentative and does not comprehensively present your results. The purpose of this step is to clarify your data.

Consider a study we conducted about role strain among professionals who served as both physical educators and athletic trainers (Pitney, Stuart, & Parker, 2008). The initial qualitative analysis revealed five emergent themes:

1. Time-related issues
2. Role relationships
3. Support and appreciation
4. Role clarification
5. Role accommodation

When we started the writing process, we discovered that the themes titled "role clarification" and "role accommodation" both represented the difficulty of negotiating dual roles. To best explain this issue, we combined the two themes into one titled, "role negotiation vs. role accommodation" which allowed for a more meaningful and accurate discussion of the findings.

Summary

The process of collecting and analyzing data to understand the perceptions and experiences of participants is the heart and soul of qualitative research. Although purposeful selection is a hallmark of qualitative research, you can select participants for your study with many different sampling strategies. Data collection and analysis occur simultaneously, but you should conduct each very systematically. You can collect data through interviews and observations, as well as by examining documents. In the process of data analysis, you attempt to identify important pieces of information that relate to the purpose of your study. Code this information and organize it to identify patterns of data.

CONTINUING YOUR EDUCATIONAL JOURNEY

 Learning Through Activity

1. Read the study by Goodwin and Compton (2004) in appendix C, focusing on the purpose statement, research questions, and researchers' interview methods. Develop a semistructured interview guide based on your observations that would give you insight about the participants' experiences.

2. Watch a live event or show on TV, assuming the role of a passive observer. Design, implement, and evaluate an observation instrument that captures the essence of human interactions in the program.

 Checking Your Knowledge

1. While planning a study, you decide that participants should have taught in a secondary public school for at least 15 years. Moreover, these educators must have

experience with community drug-awareness programs. This scenario represents which of the following sampling strategies?

a. chain sampling

b. criterion sampling

c. maximum variation sampling

d. deviant sampling

e. typical sampling

2. Over the course of many interviews, you continually hear the same sort of information. Eventually, no new information emerges. Which of the following concepts relates to this scenario?

a. triangulation of data

b. exhaustion of information

c. saturation of data

d. redundancy of data

e. c and d

3. When selecting participants for a study, you purposefully identify only one physical educator who meets your inclusion criteria, so you decide to rely on this person to connect you with other educators who fit the desired profile. This scenario represents which sampling strategy?

a. chain sampling

b. criterion sampling

c. maximum variation sampling

d. deviant sampling

e. typical sampling

4. Establishing good rapport with the participant during the initial steps of an interview is known as _____. Speaking to the participant about matters directly related to the research questions is known as _____.

a. the relational aspect; the thematic aspect

b. the thematic aspect; the relational aspect

c. authenticity; verification

d. the relational aspect; debriefing

e. none of the above

5. As part of a research study, you obtain permission to observe physical educators and students in a unit on team sports. Specifically, you hope to observe interactions among students, particularly the interactions of two who have been diagnosed with behavior disorders. In negotiating your role as observer, you agree to help organize and distribute the equipment. Once that task is done, you will simply watch the students. What form of observation does this scenario represent?

a. complete participant

b. participant observer

c. observer participant

d. complete observer

6. When analyzing data from a transcript, you identify important information that relates to your research purpose. This information is called a _____, while the label you assign to its meaning is called a _____.

   a. code; meaning unit

   b. conceptual label; meaning unit

   c. code; conceptual label

   d. meaning unit; code

   e. theme; category

Thinking About It

1. You interviewed three teachers about how they include students with disabilities in their physical education classes. When you observe these teachers at work, you notice a substantial difference between what they articulated in the interviews and what they do in the classroom. Knowing that the process of qualitative research is flexible, how would you proceed?

2. You made arrangements to speak with a department chair as part of a case study, but he has a family emergency at the time of the meeting. How would you deal with this situation?

Making a Stretch

These readings will expand your knowledge on collecting qualitative data.

Kvale, S. (1996). *InterViews: An introduction to qualitative research interviewing.* Thousand Oaks, CA: Sage.

Schostak, J. (2006). *Interviewing and representation in qualitative research.* Berkshire, England: Open University Press.

Ensuring Trustworthiness of Data

Defining Trustworthiness of Data

People continuously question the truth of professional claims. For example, when a television infomercial states that a specific device or training regimen improves abdominal strength, you may ask, "How do I know I can trust the results? Are the results even valid?" Qualitative studies inevitably prompt similar questions. Researchers doubt whether the results of an empirical study can be trusted (Merriam, 1998). Were the methods sound? Did they lead to reasonable conclusions?

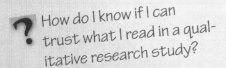 How do I know if I can trust what I read in a qualitative research study?

It is the researchers' responsibility to convince you to trust their work. View *all* research through a lens of healthy skepticism.

The quantitative measurement of variables is easily judged by the principles of validity and reliability. These traditional terms are not as compatible with the qualitative paradigm, since researchers do not attempt to measure variables. However, terms for qualitative research have been developed that closely parallel the quantitative process (Whittemore, Chase, & Mandle, 2001). Although debate still exits about how to address issues of quality in qualitative studies (Sparks, 2001; Mills, 2007), the concept of trustworthiness of data (Guba, 1981; Erlandson, Harris, Skipper, & Allen, 1993) and its components of dependability, credibility, and transferability are now standard. The principles of trustworthiness allow qualitative researchers to "...make a reasonable claim to methodological soundness" (Erlandson et al., 1993, p. 131). Therefore, you must consider issues of accuracy and quality when planning and conducting a research study if you wish your work to be taken seriously.

The following section addresses the concepts of validity and reliability. Although they do not fit the qualitative paradigm, all researchers must have a basic understanding of these principles. The discussion continues with an explanation of the various components of trustworthiness of data and their specific strategies.

Validity

Most scholars consider validity of measures the most important aspect of a quantitative research study (Ary, Cheser Jacobs, & Razavieh, 2002). Internal validity, or whether an instrument takes the intended measurement, is an important criterion for psychological variables like a person's attitude, satisfaction, or agreement in a survey, and physiological variables like the power output of a muscle, blood pressure, or blood-lactate levels.

The essence of internal validity is truth and accuracy (Silverman, 2000). From a qualitative perspective, the concept addresses whether the research findings capture what really happened and what participants truly meant and believed about a situation. Lincoln and Guba (1985) coined the term *credibility* and many qualitative researchers have since adopted it as a parallel understanding of internal validity (Merriam, 1998; Erlandson et al., 1993; Mills, 2007; Stringer, 2004).

External validity relates to how the findings of one study can be applied to other participants or settings. In other words, external validity addresses whether the results of a study can be easily generalized. Qualitative researchers do not usually employ this concept because they are more interested in understanding a specific phenomenon or situation. The findings of qualitative research studies are rarely applied to the general population. However, the findings may be applied to similar or related communities. You may read a qualitative research study and think, "Wow, this seems a lot like my physical education program," or "I have heard my students report the same thing; perhaps these findings relate to them." The extent to which qualitative findings apply to other situations is known as *transferability*.

Reliability

Reliability refers to the consistency of measures. Quantitative studies based on measurements must use an instrument that provides consistent results, regardless of the number of repetitions or the amount of rest between trials. To establish reliability, researchers often conduct two trials and compare the measured outcomes for similarities. This process is called *test-retest reliability*. Another aspect of reliability is whether results can be reproduced. If a study is repeated using the same procedures, researchers should expect similar results with a small margin of error.

Because qualitative researchers do not rely on measurements and do not perform trials of activities or conduct the same interview twice, they do not worry about whether data can be reproduced. They use the term *dependability* to denote results that are consistent with the data collected (Merriam, 1998).

Reviewing Trustworthiness Strategies

Qualitative researchers address the overarching concept of trustworthiness and the equivalents of validity and reliability (credibility, transferability, and dependability) with various strategies. Figure 5.1 outlines the questions and strategies associated with the three components of trustworthiness of data.

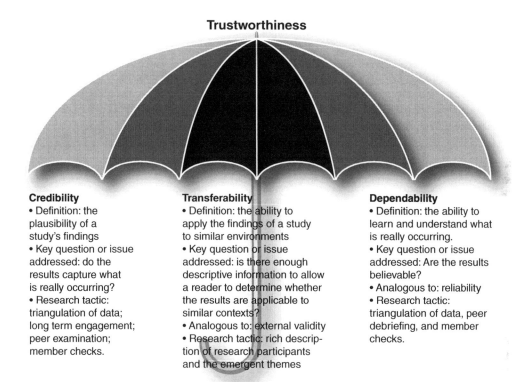

Trustworthiness

Credibility
• Definition: the plausibility of a study's findings
• Key question or issue addressed: do the results capture what is really occurring?
• Research tactic: triangulation of data; long term engagement; peer examination; member checks.

Transferability
• Definition: the ability to apply the findings of a study to similar environments
• Key question or issue addressed: is there enough descriptive information to allow a reader to determine whether the results are applicable to similar contexts?
• Analogous to: external validity
• Research tactic: rich description of research participants and the emergent themes

Dependability
• Definition: the ability to learn and understand what is really occurring.
• Key question or issue addressed: Are the results believable?
• Analogous to: reliability
• Research tactic: triangulation of data, peer debriefing, and member checks.

Figure 5.1 The umbrella of trustworthiness.

Credibility

The concept of credibility relates to whether the findings of a study are believable. Researchers must take steps to ensure that their findings are accurate and supported by the data. Credibility addresses key questions, such as, "Did the researchers depict what actually occurred in the research setting?" and "Did they learn what they intended to?" Strategies for addressing this component of trustworthiness include triangulation, prolonged collection of data, member checks, and peer reviews.

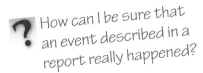

How can I be sure that an event described in a report really happened?

Again, qualitative researchers must convince readers that the events they report actually occurred. They should provide enough detail and supporting evidence to meet the standards for credibility.

Triangulation

The term *triangulation* comes from the vocabulary of navigation, in which a location is determined by its distance from two or more other points. In qualitative research, data triangulation means that your findings can be verified by other sources (Patton, 1999). Think of it as cross-checking your data.

For example, in a qualitative study of the professional socialization of athletic trainers in the NCAA Division I setting, Pitney (2006) found that athletic trainers were concerned by bureaucratic influences in the athletic setting. The ideas and beliefs of the athletic trainers were verified by participants who were athletic directors. Pitney did not simply rely on one perspective, but rather sought out other sources to confirm his findings.

Qualitative researchers also use other forms, including multiple-analyst triangulation, methodological triangulation, and theoretical triangulation (Patton, 1999). In multiple-analyst triangulation, different researchers conduct separate analyses and then compare and contrast initial findings. The meaningful themes that emerge for each analyst may vary. When this occurs, the members of the research team must negotiate how the data should be categorized. This process generates answers to research questions that pay excellent attention to detail and carefully consider all relevant data.

Methodological triangulation uses different methods of data collection. Consider a public health research project conducted by Khunti et al. (2007). In their study examining barriers to healthy lifestyles among students with type-2 diabetes, the authors conducted focus groups and observations while simultaneously administering diet evaluations and questionnaires on physical activity. This triangulation of methods provided a fuller understanding of the barriers to health perceived by the children.

Theoretical triangulation means that researchers view the data from different perspectives. For example, a researcher investigating the educational process in a general college class may interpret the data from a pedagogical perspective and then from an androgogical perspective; that is from a learning theory for both children and adults. Theoretical triangulation often helps researchers explain their findings thoroughly.

Triangulation is considered an excellent strategy for ensuring trustworthiness, especially when combined with participant checks, which are explained later in the chapter. The credibility of a study increases when researchers use appropriate methods of triangulation. Table 5.1 summarizes the various methods of triangulation.

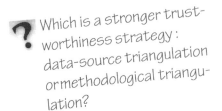
Which is a stronger trustworthiness strategy: data-source triangulation or methodological triangulation?

There is no concrete answer to this question. When examining trustworthiness of research you must consider the entire context of the study.

Prolonged Data Collection

In order to have credible findings, you may need to collect data over a long period of time. Imagine, for example, conducting a qualitative study of a physical education teacher education (PETE) program that uses a unique mentoring process for its

Table 5.1   Forms of Triangulation

FORM	DESCRIPTION	EXAMPLE	BENEFIT
Data-source triangulation	Researchers obtain multiple forms of data to cross-check information.	They interview different stakeholders (teachers, students, parents, and business owners) about an issue affecting a school district and community.	Understanding different participant perceptions helps researchers verify preliminary findings and gain a holistic perspective.
Multiple-analyst triangulation	More than one researcher analyzes qualitative data.	Three members of a research team analyze interview transcripts and discuss the emergent themes.	Data is examined thoroughly, which ensures that information from participants is interpreted appropriately.
Methodological triangulation	Researchers use more than one method of data collection.	Researchers collect data through interviews, observations, and document analysis.	Researchers verify their findings with a different method, making sure that what they observed in an interview is consistent with what they learned.
Theoretical triangulation	Researchers view the data from different theoretical perspectives.	Researchers analyze data from a study on the perceptions of students with obesity through two different lenses: the functionalist perspective and the interactionist perspective.	Different theoretical views of data yield different interpretations. For example, the functionalist perspective finds that environment influences behavior, but the interactionist perspective sees people as active participants in their world.

students. If you visit this setting once to conduct a few interviews and observations, you will likely learn a great deal, but will you capture how the program is really run? Although your classroom observations may yield good information, the students might act differently around a visitor, or feel that they need to put on a show. If you conduct observations over several weeks, the students will get used to you and are more likely to act normally.

Think of this strategy with the metaphor of photography. A picture of a lake may depict students kayaking through calm waters on a bright sunny day. A snapshot of the same shoreline on a different day may show overcast skies, choppy water, and kayakers fervently paddling to shelter. Both photographs reveal different aspects of the same activity.

Prolonged data collection yields deeper understanding, a holistic sense of the research setting, and the assurance that you have captured what really occurred.

The obvious drawback to this strategy is that you will need more time to complete your study.

Participant Checks

In participant checks, also called member checks, researchers ask the study's participants to verify the findings based on their experiences and perceptions (Pitney & Parker, 2001). Researchers conduct these checks before establishing a formal report (Mills, 2007). Participant checks are very effective, especially when used in conjunction with triangulation. Who is better qualified to verify the plausibility of research findings than the people who provided the information?

Participant checks can be performed in two ways. The first is what we call *transcript verification,* in which participants review the transcripts from their recorded interviews for accuracy. The procedure is easy to conduct. Researchers give the interviewees a copy of the transcripts and ask them to read them and make any necessary corrections.

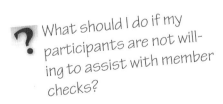 Even though I have an audio recording of the interview, my participant disagrees with the transcript. What should I do?

1. Return to the recording and check the content again.
2. If you think that the content is clear, play the recording for your participant and articulate your position.
3. Ask why your participant disagrees with the content of the transcript.

If your participant still wants to recant, you must honor that request, even if the information in question is critical to your study.

In the second strategy, called *interpretive verification,* researchers ask participants to review their interpretation of the findings. Specifically, participants receive an explanation of the study's emergent themes as well as the supporting quotes from their interviews. They can then comment on the plausibility of the research findings. This strategy allows researchers to verify that their interpretation of the data is reasonable. When performing this type of participant check, researchers should show participants how they organized the transcripts. If participants disagree with a finding, researchers must attempt to understand why. Often, participants' explanations allow researchers to reformulate their interpretations and change their results.

What should I do if my participants are not willing to assist with member checks?

Ask yourself why your participants are unable to help. Is it a conflict of time that could be resolved with scheduling, or is something else going on? The answer to this question will determine your next step. You may need to check your data and interpretations through peer debriefing instead.

Peer Debriefing

In a peer debriefing, someone with formal training and experience in qualitative research examines a study to ensure it was conducted in an appropriate and systematic manner. The reviewer, often a colleague or peer of the primary researcher, investigates six components:

1. Background information
2. Data-collection procedures
3. Data-management processes
4. Transcripts, field notes, observation summaries, and any other data
5. Data-analysis procedures
6. Research findings

The researchers first give the reviewer any background information related to the study's theoretical framework, problem statement, purpose statement, and research questions. The reviewer uses this information to become acquainted with the study's aims and methods of data collection and analysis. The researchers next explain how they collected and analyzed data, and give the reviewer any interview guides or observation forms. The goal of this component is to allow your peer to understand how data was collected and the type of data that was analyzed.

Next, the reviewer must learn how the researchers managed the data. How did they analyze their transcripts, field notes, and observation summaries? This component helps the reviewer navigate the volumes of data and make sense of the researchers' process. Each researcher has a unique system for organizing coded concepts, identifying tentative themes, and locating meaningful information.

Finally, the reviewer will audit the data for proper methods of collection, analysis, and management. For example, the reviewer examines transcripts for bias, making sure that the interviewer did not ask leading questions to obtain specific quotes. After auditing the procedures of data analysis, the reviewer looks for structure and coherence among the findings, ensuring that meaningful pieces of data were logically organized into themes.

 Who is being debriefed in this process, my peers or me?

You as the researcher are being debriefed. Debriefing brings closure to a process or an event. In this situation, you should share what you learned about the process with your peers to the extent that they can verify that what you learned is consistent with the data you collected. This process ensures that your interpretations are appropriate.

Transferability

Although quantitative researchers seek to generalize a study's findings, qualitative researchers are more concerned with the context of their participants' experiences.

However, when similarities occur between and among contexts, the results of a qualitative study may be transferred. The transferability of a qualitative study's findings is beyond the researchers' control. Ultimately, this decision belongs to a study's readers. However, researchers must provide as much information as possible when describing the context and research findings so readers can best apply the results to their particular contexts. Qualitative researchers are often at a slight disadvantage when trying to paint a picture of the research setting. It can be difficult to provide rich, descriptive information when writing manuscripts for research journals, which often have limited space.

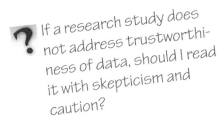 If a research study does not address trustworthiness of data, should I read it with skepticism and caution?

Yes, to a certain extent. The section on trustworthiness often is the first to be cut if a publication is pressed for space. However, a study should include at least one sentence addressing the issue.

Dependability

From a qualitative perspective, dependability relates to research processes that are clear and appropriate. Stringer (2004) states that dependability "...is achieved through an inquiry audit whereby details of the research process, including processes for defining the research problem, collecting and analyzing data, and constructing reports are made available to participants and other audiences" (p. 59). Several strategies for ensuring dependability exist, although many authors use audits to address the quality of this criteria (Lincoln & Guba, 1985; Erlandson et al., 1993; Merriam, 1998).

Creswell (1998) calls this process an "external audit," which means that someone examines the research process and product to ensure that the study's findings are consistent with its data. This process is very similar to peer debriefing, and in all likelihood the peer debriefer can also perform the dependability audit. Researchers who plan to use dependability audits should keep memos documenting the evolution of the emergent themes, their answers to research questions, any changes to interview questions, and the details of participant selection. Memos can help auditors answer many questions about the research process.

Considering the Contextual Influences

Qualitative researchers use many strategies for ensuring trustworthiness. The most frequently used strategies are triangulation, participant checks, and peer reviews. Creswell (1998) suggests that researchers use at least two strategies per study. Researchers should consider the context of their research, the purpose of the study, the study's limitations, and issues of practicality when choosing strategies. For example, if you are conducting a qualitative study and hope to understand a particular sport team's culture, it may not be practical to stay for a prolonged period of time, despite the fact that you know that would be an optimal way to fully understand the culture. This, then, becomes a limitation of your study that you must acknowledge.

Furthermore, the research context may make it impossible to conduct both interviews and observations. Participants might feel perfectly comfortable in one situation, but object to the other. Another research plan might best be served by a single case study. These situations eliminate the strategy of data triangulation.

Summary

One of the quality criteria for qualitative research is ensuring that the findings from the study are trustworthy. Trustworthiness is a general concept comprised of credibility, transferability and dependability. Trustworthiness can be addressed in many ways, but the most common strategies used are participant checks, peer review and debriefing, and triangulation. Novice researchers should use at least two strategies per study.

CONTINUING YOUR EDUCATIONAL JOURNEY

 Learning Through Activity

1. Examine the qualitative research study in appendix B and identify the steps that Pope and Sullivan (2003) used to ensure trustworthiness. Do you believe the authors did an effective job?

2. Read the study in appendix C (Goodwin and Compton, 2004). Compare and contrast their strategies with those of Pope and Sullivan.

3. Locate and read five qualitative research articles. List the strategies the authors used to ensure trustworthiness and identify which occurred most frequently. Why do you think this is?

 Checking Your Knowledge

1. When an instrument measures the intended variable, it illustrates this concept:

   a. internal validity

   b. external validity

   c. reliability

   d. triangulation

   e. a and b

2. The concept of _____ refers to research findings which can be easily generalized.

   a. internal validity

   b. external validity

   c. reliability

   d. triangulation

   e. a and b

3. This concept, a component of trustworthiness, relates to whether a study's findings are plausible.

   a. validity

   b. reliability

   c. credibility

   d. triangulation

   e. objectivity

4. This concept, another component of trustworthiness, relates to research processes that are clear and appropriate, and have findings that are consistent with the collected data.

   a. dependability

   b. transferability

   c. credibility

   d. validity

   e. none of the above

5. Which of the following examples is consistent with data-source triangulation?

   a. Two or more analysts examine the data and document their results.

   b. Researchers use different conceptual frameworks as a lens to analyze the data.

   c. Researchers conduct interviews with health educators, parents, and students to understand perceptions of obesity.

   d. Researchers both interview and observe physical educators.

   e. b and c

Thinking About It

1. You are planning a qualitative research study that examines how physicians maintain balance between work and home. How would you attempt to triangulate your data sources?

2. You show a participant the transcript of her interview as part of your participant checks. She points to a paragraph and states, "I didn't say this. This is not what I said." How would you respond to her? What steps would you take to follow up?

3. You are analyzing data for a study that is using multiple-analyst triangulation. The researchers disagree on the emergent themes. What should the research team do to remedy this issue?

Making a Stretch

These readings will expand your knowledge on trustworthiness of data.

Patton, M.Q. (1999). Enhancing the quality and credibility of qualitative analysis. *Health Services Research, 34,* 1189-1208.

Pitney, W.A. (2004). Strategies for establishing trustworthiness in qualitative research. *Athletic Therapy Today, 9*(1), 45-47.

Trochim, W.M.K. (2006). *Research methods knowledge base.* Web Center for Social Research Methods: www.socialresearchmethods.net/kb/qualval.php

Attending to Ethical Issues

An Overview of Research Ethics

Research ethics have received a great deal of attention in the recent past. In response to federal regulations, both universities and private businesses have increased rules for researchers who use human participants. This chapter first explains why research is regulated and identifies the underlying principles of research ethics, including the process of informed consent. It next articulates the role of institutional review boards (IRBs) in protecting human participants. Finally, it highlights problems that may arise during research and the steps researchers must take to protect their participants.

Research Regulations

Government guidelines, regulations, and requirements are in place because researchers have flagrantly harmed and abused human participants in the past (Quinn, 2004). Examples of these unfortunate historical events include the Tuskegee syphilis study, the Jewish Chronic Disease Hospital study, and the Willowbrook study (Noble-Adams, 1999). Though the specific circumstances differ for each of these studies, they all involve ethical violations of participants, such as intentionally infecting them with diseases, coercing them into a study, failing to inform them of a study's risks, or experimenting on them without their knowledge.

Principles of Research Ethics

The theoretical basis of ethics, or moral philosophy, is the foundation of the underlying principles that guide research (Aita & Richer, 2005). Two fundamental theories exist: deontological and teleological ethics. Deontological ethics evaluate the merit of individual actions, regardless of their outcome. For example, if researchers are dishonest to participants about the nature of a study, they violate deontological ethics because honesty is a core value of the research society. Teleological ethics address whether the outcome of a study is good or bad. Does the end justify the means? In the preceding example, the researchers may believe that the outcomes of their study will excuse their dishonest conduct with the participants. Perhaps deceiving participants will lead to remarkable findings and contribute to the good of many people (Aita & Richer, 2005).

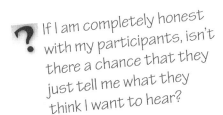 If I am completely honest with my participants, isn't there a chance that they just tell me what they think I want to hear?

This situation is possible. However, if you carefully state the purpose of your study in non-judgmental terms, your participants should feel comfortable sharing their perspectives.

General principles have emerged from these ethical foundations to guide researchers' actions when developing studies and collecting data from participants. The most common principles were generated by reports of massive abuse, such as the Nuremberg code from World War II and the Belmont report from the National Commission for the Protection of Human Subjects of Biomedical and Behavioral Research (Jeffers, 2002). The following section elaborates on these general principles:

1. Respect for autonomy
2. Beneficence (nonmalfeasance)
3. Respect for dignity
4. Justice

Respect for Autonomy

Researchers protect participants' autonomy by making sure they voluntarily consent to be involved in a study. If participants choose not to be involved, researchers must absolutely respect their decision. Researchers may never threaten people with negative consequences if they do not participate in a study.

Researchers must provide enough information about the study for potential participants to make a rational decision whether to be involved (Aita & Richer, 2005). This concept, called *informed consent,* is another element of respect for participant autonomy. Informed consent is often provided in writing.

Composing the Informed Consent Form

The informed consent form explains the following elements of the study:

- Its purpose and benefits
- Any possible risks of participation
- Clearly delineated expectations of the participant
- The voluntary nature of participation
- The participant's right to withdraw without prejudice
- How the participant's information will be treated
- The participant's right to ask questions

The informed consent form should be written in language that is easy for participants to understand. Remember, your participants may not be experts in a given subject or have a full understanding of your theoretical framework. Clearly articulate the purpose of your study and what you hope to learn. Explain how the findings of the study will inform current practice or knowledge and benefit others, such as students, teachers, patients, organizations, or professionals. Emphasize that participation is voluntary and that participants can withdraw at any time without consequence. Explain exactly what the participant will do or experience. If you plan to conduct personal interviews, indicate the location, approximate length, and number of interviews. Outline similar details about any intended observations.

 One of my participants withdrew from the study after I collected and analyzed all of the data. Do I need to analyze my entire study again or can I keep the participant's data?

Talk to your participant about the situation, but you must respect his or her desire to remove any data.

In the section of the form about treatment of data, explain that you may present or publish your findings in a report. Describe what you plan to do with recordings of interviews and summaries of observations after the study is finished. Finally, you must explain how you plan to maintain the participants' privacy. The main methods of identity protection for participants are anonymity or confidentiality.

Anonymity is rare in qualitative research because it means that researchers do not know the identity of any of the participants. A qualitative researcher could obtain anonymous data from participants by conducting a survey through the mail. The principle of confidentiality means that researchers know who their participants are, but take care to conceal their identities. Qualitative researchers commonly address confidentiality by replacing their participants' names with pseudonyms in transcripts and reports. The informed consent forms with the participants' real names are only seen by the researchers and are locked up safely and never disclosed during or after the study is complete.

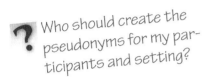 Who should create the pseudonyms for my participants and setting?

The pseudonyms must appropriately reflect the culture of the participants and setting, and either you or your participants may select them.

Participants must understand that any questions they have will be answered. Provide contact information for members of the research team in case participants have questions about the study itself. If participants have questions about their rights, consider providing contact information for a higher authority, such as a faculty advisor, a school official, or an organizational leader.

The informed consent form is a critical part of qualitative research, even if the study poses little or no perceived risk for participants. The use of informed consent is required by federal regulation and is overseen by the Office of Human Research Protections (OHRP). The OHRP's Division of Compliance Oversight reviews institutions' (i.e. colleges and universities) procedures for ensuring that participants are protected and ethical principles are not violated. If researchers want to conduct research, but they are not associated with a college or university, they are still expected to follow the guidelines from the Declaration of Helsinki that explicates research ethics guidelines. See the "making a stretch" section below for a link to information related to the Declaration of Helsinki.

Participants are required to read and sign the consent form. Provide a separate signature line if you plan to record your interviews. Researchers often give participants a copy of the form for their records. Figure 6.1 shows an example of an informed consent form.

Obtaining Full Consent and Assent

In some studies, it is nearly impossible for researchers to obtain written consent. Other times, consent is implied. In studies based on surveys, researchers use cover letters to explain the study, their expectations of participants, and how they plan to handle returned surveys. Voluntary participation is implied by those who return the survey.

Qualitative researchers who conduct interviews should obtain both written and verbal consent. We recommend asking participants for permission to audio record the conversation before you start the interview. In a personal interview, begin recording before you ask for permission so you capture the statement of verbal consent if it is given. Turn the device off if the participant does not wish to be recorded. However, according to Federal Communication Commission (FCC) guidelines, you may not record a phone conversation until you have obtained the participant's consent.

The Prevalence of Role Strain Among Dual-Position Physical Educators and Athletic Trainers in the High School Setting

Participant Consent Form

A. Authorization

I, (participant's name)_____, hereby consent to participate in a study which will involve interviews performed by (researcher's name) _____.

B. Description

- This study will involve a personal interview about my past and current experiences as a physical educator and athletic trainer in the high school setting. The study is being conducted through Northern Illinois University.

- The interview will take 30-45 minutes of my time.

- The purpose of the study is to gain a better understanding of role strain among those who are both physical educators and athletic trainers in the high school setting. The study attempts to understand how individuals have experienced their professional roles in the high school setting.

- As a participant, I will be interviewed and asked several questions related to my experiences as an athletic trainer and physical educator. The interview will be audio recorded. Although the questions are not intended to be sensitive in nature, I may elect to refrain from answering questions if I so desire. Additionally, I may terminate the interview at any time without risk of prejudice or penalty.

- There are no experimental procedures or physical risks involved with this study. There are no perceived emotional, social, or psychological risks. My name will be kept confidential and will only be divulged among the researchers. It will not be divulged in a verbal or written manner. If I request, the researchers will not include any specified information in a research report.

- If I am quoted in any way on a research report, I will be given a pseudonym. The same is true of any other individuals, institutions, or organizations that I mention in the interview.

- Once the audio files are transcribed, the researchers will modify and code the names from the transcripts to secure anonymity. Once the study is completed, the audio files and coding sheet will be destroyed.

- If I have any questions about the research study, or if I have any question about my rights as a research participant, I can contact:

  (Researcher name)_____

  (Office of Research Compliance)_____

  (Department, university, city)_____

  (Research university)_____

  (Other contact information)_____

  (Contact information)_____

»continued

Figure 6.1 Example of an informed consent form.

»continued

C. Benefits

The researchers believe that understanding the process of role strain involved with dual-position physical educators and certified athletic trainers may help improve the professional development of these individuals.

D. Voluntary Participation

I understand that participation is voluntary and that I will not be penalized if I choose not to participate. I also understand that I am free to withdraw my consent and end my participation in this project at any time without penalty by notifying the project director,

_____.

E. Consent

I have read and fully understand the consent form. I sign it freely and voluntarily. I have received a copy of this form.

Date: _____   Time: _____ (a.m./p.m.)

Signature of participant: _____

Signature of witness: _____

I agree to be audio recorded during the interview.

Signature of participant: _____

Once you have signed the consent form, please place it in the postage-paid, preaddressed envelope provided. Remember to keep the other copy of this form for your records.

Figure 6.1   *continued*

If you plan to interview minors (people younger than 18 years), you must first obtain consent from their parents or legal guardians. If the legal guardians agree to let their children participate in the study, you must next obtain written assent from the minors themselves.

 What should I do if the parent or guardian gives consent but the minor does not want to participate in a study? Similarly, what should I do if a minor wants to participate but the parent or guardian does not consent?

You may not include the minor in your study in either case.

Beneficence

The principle of beneficence, also called *nonmalfeasance*, directs researchers to respect the welfare of participants and do no harm. At the very least, researchers must attempt to balance the possible risks of a study with its benefits (Aita & Richer, 2005; Noble-

Adams, 1999). Forms of risk for participants include psychological, emotional, physical, and social harm.

Researchers may cause psychological or emotional harm if the study makes the participants experience stress or anxiety. For example, if you are conducting a study of how athletes cope with the death of a teammate, your interview questions may uncover emotions that the athletes have not processed. They may experience a great deal of stress while discussing their reactions to the death.

It is easy to understand how research of professions dealing with health and physical activity may cause physical harm. Tests of physical activity carry inherent risks as participants may sprain an ankle or strain a muscle. Observing a participant's natural behavior, conducting an interview, or examining a participant's written work generates very few physical injuries. Therefore, in most cases qualitative research presents almost no risk of physical harm.

We use the term social harm to capture instances where a participant may inadvertently suffer isolation, marginalization, or disaffection due to their involvement in a research study. One way in which these can arise is if you, as a researcher, have a power relationship over someone who is a participant in your study. In such instances there is risk of harm because if he/she participates in your study you may learn something about them that causes you to have prejudice against them. Another potential cause of social harm is instances where an informant's identity is not protected. That is, an interviewee's identity is inadvertently made known to others. If an interviewee provided discriminating information about others, he may find himself in a difficult situation or predicament. Researchers should take measures to maintain confidentiality of their participants by replacing their names with pseudonyms.

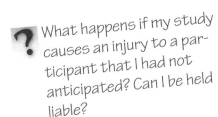 Can I identify my participants to my advisor? If not, how can I demonstrate that I have obtained an adequate pool of participants?

You should be able to explain the range of participants to your advisor without revealing specific names.

Qualitative studies are rarely so sensitive that the process harms the participants, but the potential for injury can be difficult to predict (Noble-Adams, 1999). We recommend that you discuss your study with several colleagues, or with other researchers who have conducted similar studies, to identify any potential risks. You must clearly indicate the possibility of harm in the informed consent form.

What happens if my study causes an injury to a participant that I had not anticipated? Can I be held liable?

We believe that under Tort law you or your institution can be held liable if there is evidence that all four of the following components were present: duty, breach of duty, proximate cause, and injury.

Dignity and Justice

As Noble-Adams (1999) discussed, the respect of participant's dignity relates to not using coercive methods to recruit participants and ensuring that participants are treated in a courteous and respectful manner at all times. Any agreements made between a researcher and participant at the beginning of a study, or at any time during a study, should always be upheld; there is no place for deceptive actions during your research (Mills, 2007).

The principle of justice, or fairness, requires researchers to treat all participants equitably during the research process. For example, if you advertise an incentive, such as a small gift or gift certificate, for participants who consent to interviews, you must make it available to all the interviewees.

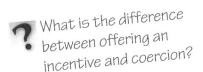 What is the difference between offering an incentive and coercion?

An incentive is usually a small gift or chance to win a small token of appreciation. Coercion is more sinister and suggests a punitive or negative consequence for non-participation.

Special Considerations

Researchers must be aware of special circumstances before beginning a study, such as working with vulnerable participants and identifying situations which may require the disclosure of a participant's involvement and identity.

Respecting Vulnerable Participants

In some studies, you may need to collect data from participants who are part of a population considered vulnerable. Examples include minors (children), prisoners, people confined in detention centers, and people with physical or mental disabilities. Researchers who include vulnerable populations in their studies must take special care to ensure that they do not exploit their participants. Usually researchers must obtain special permission from the participants' parents, guardians, or administrative supervisors before collecting data.

Disclosing a Participant's Identity

In general, confidentiality dictates the protection of participants' identities. Although they are extremely rare, some instances necessitate disclosing a participant's identity.

You must disclose the identity of participants who indicate that they intend to harm themselves or others. Consider the previous example of the study about how athletes cope with the death of a teammate. During the course of an interview, a participant begins to cry and states that he can't handle his sorrow. He shares suicidal thoughts with you and threatens to harm himself. Now, you must intervene by reporting this information immediately to the appropriate contacts.

You may also disclose a participant's identity in cases of suspected abuse. Consider a study in which you are interviewing children about their physical activity at school and home, you have obtained both parental consent and participant assent to conduct interviews. A child tells you during an interview that her parent strikes her and burns her with cigarettes. You observe bruises on the child's face and burn marks on her arm that support the allegation of abuse. You are now required to take action by reporting this information to the appropriate authorities.

You must prepare for these rare instances by understanding your responsibilities and fully inform your participants of actions you may be required to take. Consider adding a statement to your informed consent form that identifies your obligation to intervene, such as, "In case of any threat to yourself or others, your name will be provided to authorities."

Institutional Reviews

If you are conducting a study to fulfill requirements for a graduate degree, you will probably need to submit your proposal to an institutional review board (IRB). Federal law requires the critical analysis of student proposals by such boards. IRBs pay particular attention to the general principles discussed earlier in this chapter, and often require you to complete a form with specific questions like these:

- What is the study's purpose?
- How will you recruit participants?
- What will you require participants to do?
- How will data be collected and managed? Will it be stored in a secure location? Who will have access?
- What are potential risks of participation in this study? How will you minimize potential harm?
- Will any support services be available to participants?
- What are the proposed benefits of the study?
- Will the study involve members of vulnerable populations? If so, which ones?
- How will you obtain informed consent?
- Have members of the research team had training in research ethics?

In addition to answering these questions, researchers must provide copies of all documents for review, such as consent forms, cover letters, and interview questions. The IRB panel may ask researchers to clarify or change their procedures if there are undue risks to participants.

The discussion of research ethics in this chapter has related to the stages of planning, data collection, and data analysis. However, ethical considerations also apply to the process of writing and publishing research. Remember to protect the privacy of your participants in research reports and publications. You must fairly portray the meanings and expressions of your participants when writing your results. Make honest interpretations based on existing data, not data you wish you had collected. Represent your sources accurately and appropriately, giving credit where credit is due.

Summary

Unfortunate historical events have generated many research regulations that are designed to protect human participants from harm. Researchers must be aware of the ethical principles that guide their actions, including respect for autonomy, beneficence, respect for dignity, and justice. Informed consent, usually in writing, is a universal procedure that helps participants fully understand their involvement in a study and how their data will be treated. The institutional review of student research proposals is mandated at colleges and universities to protect human participants and ensure that ethical principles are followed.

CONTINUING YOUR EDUCATIONAL JOURNEY

 Learning Through Activity

1. Find and read three published qualitative research studies. Identify potential risks that the authors may have considered while conducting their studies.

2. Revisit the study by Pitney (2002) in appendix A. Design an informed consent form for use with this study.

3. Ask qualitative researchers how they store transcripts, informed consent forms, audio recordings, and other study materials. Also, ask them how they dealt with ethical dilemmas during their research studies.

 Checking Your Knowledge

1. Deontological ethics refers to _____. Teleological ethics refers to _____.
   a. whether the outcome of an action is good or bad; whether an action is right or wrong, regardless of the outcome
   b. whether an action is right or wrong, regardless of the outcome; whether the outcome of an action is good or bad
   c. the culturally sensitive aspects of a social interaction; whether the outcome of an action is good or bad
   d. whether an action is right or wrong, regardless of the outcome; the culturally sensitive aspects of a social interaction

2. When you ensure that participants voluntarily enter your study, you are following which principle of research ethics?
   a. respect for autonomy
   b. beneficence
   c. justice
   d. respect for dignity
   e. b and c

3. When you know your participants' names but take measures to conceal their identities, you are following which principle of research ethics?

a. disclosure

b. autonomy

c. anonymity

d. confidentiality

e. c and d

4. Which of the following principles requires researchers to do no harm to participants, or at least to balance potential harm with potential benefits?

a. beneficence

b. respect for dignity

c. nonmalfeasance

d. a and b

e. a and c

5. Which of the following participant populations are considered vulnerable?

a. college professors

b. students in a juvenile detention center

c. patients with Alzheimer's disease

d. b and c

e. a and b

Thinking About It

1. While conducting interviews at a junior high school, a student participant tells you that she has begun to self-inflict wounds to her arms. As you glance at her arms, you see fresh cuts that seem to confirm her comment. What should you do? Whom would you contact?

2. A health educator you are interviewing asks you to shut off your recording device and discuss a matter off the record. How would you proceed?

3. You have interviewed and observed a physical therapist for a study. She tells you that she does not have time for the follow-up interview you have scheduled and that she wishes to drop out of the study. What is the most prudent response?

Making a Stretch

The following book and tutorial will expand your knowledge of research ethics.

Stanley, B.H., Sieber, J.E., & Melton, G.B. (1996). *Research ethics: A psychological approach.* Lincoln, NE: University of Nebraska Press.

The National Institutes of Health's (NIH) office of extramural research hosts a Web-based tutorial for protecting human participants of research studies. Visit the site at http://phrp.nihtraining.com/users/login.php to learn more about research ethics. When you successfully complete the quizzes at the end of the online tutorial, you receive a certificate of completion.

The British Medical Journal has published information about the Declaration of Helsinki. Visit this information at http://www.cirp.org/library/ethics/helsinki/ to learn more about the research principles included in this Declaration.

Writing Qualitative Research

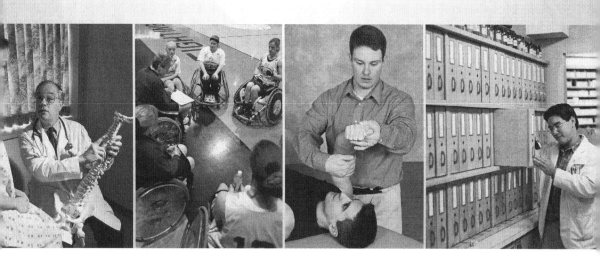

Part III begins with chapter 7, which explains how to write research proposals for obtaining both approval for academic studies and funding through small grants. The chapter focuses on the components of a research proposal, namely the introduction, literature review, and methods sections.

For chapter 8, we assume that you have conducted a qualitative research study and are now ready to write a research report. This chapter outlines ways to present your results using participants' quotes to support your findings. It also provides advice about how to write the discussion and conclusion sections.

Guiding Questions

Consider the following questions before reading part III. They will guide your examination of each chapter.

1. Why is a research proposal necessary?
2. What is the difference between writing a proposal and writing a research report?
3. What is the difference between an academic proposal and a grant proposal?
4. What options do you have when seeking a balance between the voices of participants and researchers in your results section?
5. How can you use participants' quotes to support a study's thematic findings?

The following figure illustrates the content, connections, and organizational configuration of part III.

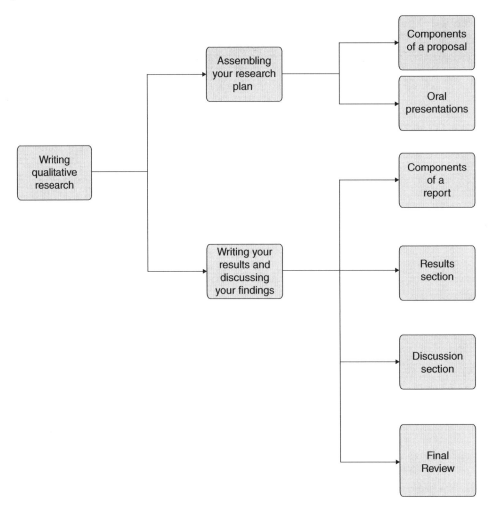

7

Assembling Your Research Plan

Learning Objectives

Readers will be able to do the following:

1. Explain the process of creating a research proposal.
2. Identify the components of a research proposal and distinguish among grant, thesis, and dissertation proposals.
3. Explain why a thorough proposal is important.
4. Create a research proposal.

Components of a Proposal

A research proposal contains several critical components (or chapters), including the introduction, literature review, and methods, and can take various forms depending on the audience and the author's intent. For example, a proposal for a thesis or dissertation differs from a grant proposal, which is tailored for the requirements of the funding agency. This chapter explains each component of a research proposal, and offers examples that illustrate the various forms. It also explains the difference between an academic proposal, which leads to a thesis or dissertation, and a grant proposal, which seeks funding for a study. The section on grant proposals focuses on small grants that are internal or local rather than larger ones that are national or international.

Before reading this chapter, you should have identified a topic and research questions for your study, as well as started thinking about the study's significance. Moreover, you should have begun consulting literature and identifying appropriate strategies for collecting and analyzing data. If this is not the case, please read chapters 3 through 6. This chapter helps you format these critical aspects of research for either an academic research proposal (thesis or dissertation) or an application for funding (grant).

Table 7.1 illustrates the critical components of a research proposal and outlines the similarities and differences between academic and grant proposals. While Part II presented these components from the standpoint of conceptualizing and planning a study,

Selected text on pages 89-93 from *An investigation of the social support network of injured athletes* by Kari M. Borseth Glenn. Unpublished master's thesis at NIU, DeKalb, IL. Reprinted by permission of Kari M. Borseth Glenn.

Table 7.1   Critical Components: Similarities and Differences Between Academic Proposals and Grant Proposals

COMPONENT	ACADEMIC PROPOSAL (THESIS OR DISSERTATION)	GRANT PROPOSAL
Introduction	• Conceptual framework • Statement of the problem • Purpose statement • Research questions • Significance of the study • Author's perspective • Definition of terms	• Conceptual framework • Statement of the problem • Purpose statement • Research questions • Significance of the study • Links to mission of funding agency
Literature review	Should be extensive, covering critical articles and presenting interpretations to support current study	Often a short, condensed version of an extensive review that includes critical articles to support current research
Methods	• Participants • Sampling • Procedures • Data collection • Data analysis • Ensuring trustworthiness	• Participants • Sampling • Procedures • Data collection • Data analysis • Ensuring trustworthiness • Anticipated outcomes (where will results be disseminated?)
Timeline	Needed for plans to complete the study (helps establish feasibility with committee members)	Needed to assure funding agency that you will adhere to deadlines and complete the project in a timely manner
Budget		Must be clear and items must match the funding available from the agency (always check carefully to determine whether items such as travel, conference registrations, equipment, etc. can be funded)
Support/ matching funds		If possible, show evidence of matching funds or supplemental funding
References	Must be included to give adequate credit to literature that has guided your conceptualization of the study	Must be included to give adequate credit to literature that has guided your conceptualization of the study
Appendices	• Informed consent • Observation/interview guides	• Informed consent • Observation/interview guides • Letters from advisor/dean/chair • Obtain letters of support from critical parties (advisor/chair/dean)

this section focuses on pulling all of these components together into a coherent product that allows you to share your research plan with others.

Introduction

As we mentioned in Chapter 3, the introduction should ease readers into your topic and lead them logically toward your purpose statement and research questions. In short, it provides a context for your research that allows readers to understand what you intend to accomplish and why your study will be significant, or how it will contribute to the existing body of knowledge. The introduction should follow a logical progression to outline the course of your study.

 To whom should I address the introduction?

Are you writing a proposal or an article? Write the introduction of a proposal for an interested but uninformed audience. When writing an article, consider the readership of the specific journal.

Your introduction should contain background information to introduce the topic and provide an understanding of the problem. Explain the purpose statement as a logical extension of the problem that leads to specific research questions. Once you have clearly articulated these three components, explain the significance of your study. Recall that significance may be related to policy, practice, or existing theory. When writing a grant proposal, be sure to make a direct connection between the significance of your study and the mission of the funding agency. Consider a grant proposal that Parker and Pitney wrote to a university in which they cited the link between their study and the university's mission statement:

Links to Mission of the University, College, and Department

As the proposed study will examine professional programs within the entire university, it seems appropriate to address its importance in relation to the university, college, and department mission statements. At the university level, this university is committed to "preparing students for effective, responsible, and articulate membership in the complex society in which they live as well as in their chosen professions or occupations" (Mission Statement, p.1). As previously mentioned, in our society obesity is increasing and graduates will have greater interactions with students/clients who are obese. It is, therefore, vital that our preparation programs provide students with the necessary experiences and tools to be effective professionals in their chosen field. (2004)

Clarifying Your Stance

Conducting qualitative research is a very personal process. Because you will serve as the instrument for both data collection and analysis, you should share your perspectives and perceptions when presenting a proposal. In other words, you may bring a perspective to the research process that has the potential to create bias. Therefore, you should explain your feelings, attitudes, dispositions, and general thoughts about a given topic before you begin collecting and analyzing data.

Take time to search your beliefs and write out why the topic is important to you, what you think you know about it, and what experiences you have with the topic. In a sense, you need to come clean about your thoughts on a given phenomenon before you can gain true, unbiased insight from the experiences of others. An exercise like this can be helpful for every type of qualitative study. Becoming aware of your beliefs will help you keep them out of the process of data analysis. This process will also prepare you to answer the two questions asked in many proposal meetings: "Why are you interested in this topic? What experience do you have with this topic?" Although you may not include this section of writing in your proposal document, it will help you clearly articulate your stance.

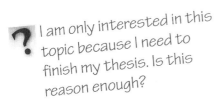 I am only interested in this topic because I need to finish my thesis. Is this reason enough?

We don't believe it is. A conscious commitment to qualitative research requires time and energy from you, your advisor, the participants, and the committee. "Just because I need to finish" does not give the research and everyone involved the respect they all deserve.

Defining Terms

As you develop your proposal, you will likely use unique terms in your introduction. Many researchers create a section of their proposal that defines special terms for readers. This section is much like the glossary of a textbook. Readers can consult it for clarification on a term or concept. It can also help you address questions about terms at your proposal meeting.

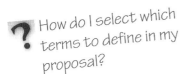 How do I select which terms to define in my proposal?

First, think about the terms that you struggled with as you developed the proposal. Next, ask an uninformed but interested party to read your introduction and tell you which parts they don't understand.

Literature Review

A literature review synthesizes, compares, and contrasts information from reliable sources. This section is your chance to show that you are familiar with existing literature related to your topic. Do not simply provide an annotated bibliography or a string of abstracts that you have written for each article or text you reviewed. This sort of review often seems like a laundry list of references with small bits of narrative about your findings intertwined. Although it is important to include your references and findings, you must work concepts together into a rich narrative.

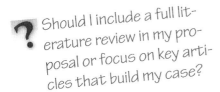 Should I include a full literature review in my proposal or focus on key articles that build my case?

An academic proposal should include a full literature review. A grant proposal should focus on key pieces.

Organize your review of the literature with the themes you identified from critical articles and texts. This type of organization shows that you have taken ownership of your interpretations and can articulate them when questioned. Check the structure of your review by examining the opening statements of each section or paragraph. If most of those sentences start with an author's name or date, you may be guilty of constructing a string of pearls, which lists the literature with few interpretations or thematic connections.

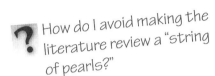 How do I avoid making the literature review a "string of pearls?"

Everyone approaches this task differently. It may be helpful to type up the key points of your articles and then sort them into categories linked by common themes. Lead with the themes in your writing rather than the articles themselves.

Before you present the themes of your literature, show readers how you intend to organize the review so they know what to expect. The following example outlines the structure of its material.

This section will present a review of current literature regarding research in the area of social support network, with the intention of specifically reviewing the coach as a main source of social support, the effect of coaches' behaviors and feedback, and the role of perceptions of coaches. After addressing the coaches' behaviors and feedback, literature on the social support network is presented.

Coaches Behaviors and Feedback as a Means of Social Support

Coaches' behaviors and feedback have a significant impact on an athlete's belief of oneself. Coaches are also considered an athlete's main means of social support (Allen & Howe, 1998; Amorose & Horn, 2000; Black & Weiss, 1992; Horn, 1984; Kenow & Williams, 1999). It has been suggested that praise and encouragement from coaches results in enhanced perceptions of physical competence, positive affect, and a desire to continue to participate in sports and improve physical skills. It has also been studied that leadership style has an influence on athlete's behaviors. These areas have been significantly studied with non-injured athletes; however, when addressing injured athletes, research is sparse. As such, investigation into the importance of and effects of coaches' behaviors and feedback on the injured athlete is warranted.

Support for the belief that coaches play an integral role in athletes' perception of ability is evidenced in many studies. One such study, conducted by Allen and Howe (1998), was designed with the purpose of examining the influence of ability and coaches' verbal and nonverbal behaviors on adolescent female athletes' perceptions of competence and affective responses to their sport participation. Allen and Howe (1998) hypothesized that adolescents with greater ability levels who perceived their coaches as giving more praise, information, and praise combined with information in response to performance and effort would report more positive affect and physical competence than adolescents with lower ability levels that received less positive responses to a performance and effort. They also believe that athletes with greater ability who perceived more encouragement, corrective information, encouragement with corrective information, and less criticism would report greater positive affect and physical competence compared with those of lower ability who perceived less encouraging information and more critical feedback in response to mistakes. (Borseth, 2004, p. 13-14)

Note that the anticipatory paragraph from the preceding example presents the structure of the review and the centralized heading indicates the first theme. The author summarizes the literature without creating a string of pearls.

Some qualitative researchers believe that reviewing the literature too much can taint your analysis of the data collected later. If you read all or most of the information related to a topic before starting your own exploration, you may not make any new discoveries. This point is both interesting and important. A certain paradox does exist, because you must clearly articulate a research plan in order to navigate the proposal process for a thesis, dissertation, or research grant. You cannot accomplish this without a literature review. However, be wary of going so far with a literature review that you are unable to engage in a true process of discovery and insight that is the hallmark of qualitative research. Look for gaps in the literature that support the need for your study. Review enough sources to effectively present your proposal, but limit the amount of information you consider.

Methods

The methods section of your proposal should address the following aspects of your study:

- Who will participate? How will you select the participants?
- What procedures will you use to identify and recruit participants? What will you require them to do?
- Where will you collect data? What will you collect and why? How do you plan to analyze the data?
- How will you ensure the trustworthiness of your data?

Describing Participants

You must describe your participants as well as you can before meeting them. The following questions should help you write this section:

- Why are these particular participants important?
- How did you select them?
- What kind of sampling will you use? (Refer to chapter 4 for a review of sampling procedures)
- How many participants do you plan to use?

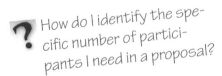 How do I identify the specific number of participants I need in a proposal?

Examine your purpose statement and, perhaps in consultation with your advisor, determine the fewest number of participants needed to address your purpose statement. List this number in the proposal, but state that the final number of participants will be determined when data saturation has been reached. See chapter 4 for a review of data saturation.

Explaining Procedures

Now that you have described the participants, you will explain in the procedures section exactly how you plan to recruit them. For example, how will you initiate your snowball sampling? You should also discuss how you will gain access to the research setting. Suppose you plan to study the experiences of students who are obese in physical education classes. How would you approach teachers or administrators and convince them to let you interview students?

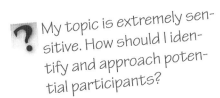 My topic is extremely sensitive. How should I identify and approach potential participants?

Why is this situation sensitive? Is the topic itself sensitive, or are you worried that participants may not want to disclose that they are part of the group or phenomenon that you wish to study? Discuss your situation with colleagues and brainstorm how to best recruit participants.

Collecting and Analyzing Data

The section of a proposal about collecting and analyzing data is critical. Be sure to share your intended methods. You must articulate how you plan to obtain information that will help you achieve your research purpose. Explain whether you will conduct interviews, perform observations, or examine documents or any combinations of these. You must also justify your choice of methods. Finally, if you plan to conduct interviews or observations, explain how you will record the data.

After collecting data, how will you analyze it? Chapter 4 explains that the steps of data analysis should be very clear and systematic. Inductive analysis, or the process of deriving conclusions from systematic evidence, is at the heart of qualitative research. In your proposal be sure to clearly outline the process as illustrated here by Pitney and Ehlers in their grant proposal for the study published in 2004:

Data Collection and Analysis

Both researchers will conduct interviews using a semi-structured technique. The interviews will be recorded, transcribed, and then analyzed inductively. The interviewers will both be involved with the initial 5 interviews to gain a familiarity with the question posing procedures to enhance the consistency of questioning.

The inductive analysis will follow grounded theory procedures identified by Glaser and Strauss (1967) and later explicated by Strauss and Corbin (1990). Grounded theory is a systematic approach for the collection and analysis of qualitative data for the purpose of generating explanations that furthers the understanding of social and psychological phenomena (Chenitz & Swanson, 1986). The grounded theory approach consists of identifying specific concepts in the transcripts that explain and give meaning to the phenomenon of mentoring. Concepts will be labeled and then organized into like categories. Both researchers will conduct the analysis together using qualitative data analysis software, specifically the *Ethnograph*™. This software allows researchers to identify concepts from the transcripts and organize them into like categories. The software also allows research memos to be entered.

Ensuring Trustworthiness of Data

You must take steps to ensure the quality of your research. Traditional forms of research measure quality in terms of validity and reliability. Because the nature of qualitative research is different, these terms are not appropriate. We explored issues of trustworthiness in Chapter 5 so suffice it to say that you will need to comment on this for your proposal and clearly indicate what specific strategies you will employ in order to conduct a study that is credible. For example, in planning a study, Pitney and Ehlers (2004) addressed trustworthiness in their proposal in the following manner:

Trustworthiness

To enhance the quality and credibility of the study, data source triangulation, a peer debriefing, and member checks will be performed (Patton, 1990, 1999). Data source triangulation (including both the protégés and mentors as participants) will be completed to compare alternative perspectives and expose any inconsistencies. The peer debriefing will be accomplished by having an experienced qualitative researcher examine the transcripts and coding sheets (which explain the emerging theme(s) as well as categories and concepts) for plausibility. Member checks will be conducted by having a minimum of three participants examine the findings to ensure that they are consistent with their experiences.

Stating Anticipated Outcomes

Funding agencies often require researchers to anticipate the outcomes of their study in their grant proposal. To be clear, you do not need to predict the results, but you must be able to articulate how you intend to disseminate them. Will you present your findings at a conference, submit a manuscript to a specific journal, or integrate the findings into your teaching?

Additional Content

Depending on the content, you may need to include additional information. Some proposals contain a limitations section that outlines any restrictions or compromises the researcher must make. Acknowledge any limitations of your study up front: Explain why they will occur, how they will affect the data, and how you intend to address them. For example, in a proposal to investigate the social support systems of injured athletes, Borseth (2004) indicated one limitation of her study would be that it focused on four sports teams and all of the sports were team sports and not individual sports. She cautions that athletes participating in individual sports may feel differently about needing a support system to assist them through injury. By addressing this issue in the proposal, Borseth (2004) also illustrates a delimitation of the study and assists the reader in determining whether her results will be transferable to the reader's own context. We must also be clear that not all limitations may be known at the time of writing the proposal. All you can be held to as the researcher is an honest and insightful description of the limitations as you see them during the planning phase of your study. In our experience, the limitations section in the final research report is sometimes much longer than the one in the proposal! Consider an example from a published report by Ball, Salmon, Giles-Corti, and Crawford on the limitations of their study on the socioeconomic differences in physical activity:

> For a qualitative investigation, the sample size in this study was relatively large. However, limitations include the possibility of socially desirable responding, although we believe this was minimal in this study, since many women gave detailed accounts of their physical activity that frequently did not match levels generally recommended for health benefits. The sample was recruited from within only three suburbs of the Melbourne metropolitan area, and so the generalizability of findings is unknown but may be limited, particularly given that neighbourhoods of similar SES may be very different. (2006, p. 111)

In addition to the main body of the proposal, you should include various appendices for all reviewers, containing such items as interview guides, informed consent forms, and observation instruments. Placing each of these in a separate appendix lets you share the information without substantially interrupting the flow of your writing. If you are writing a grant proposal, your appendices may also include letters of support from critical parties (department chair, advisor, or dean) and evidence of supplemental funding.

We suggest including an appendix with a time line for completion of your study, which is beneficial even when it is not required. The time line serves several key purposes. First, it shows that you have thought through each step of the research process and determined a realistic time frame for completion. Second, it allows the members of your committee to check their availability. For example, if committee members are scheduled for a sabbatical at a time when you are most likely to need their expertise, you must begin rescheduling or looking for a replacement. Third, if your proposal is for funding, the time line will help the funding agency ensure that your study will meet the dissemination deadlines often associated with grants.

When creating your time line, remember to invite Mr. Murphy, lest he appear as an unwelcome guest! Locke, Spirduso, and Silverman (2000) capture this concept when they state, "Murphy's Law dictates that, in the conduct of research, if anything can go wrong, it probably will" (p. 78). In other words, with careful consideration in the proposal stage, you should be able to anticipate most of the problems that could arise in the course of your study. We suggest using a teaching strategy called *if-then* (Tjeerdsma, 1995). Examine each step of your research process and try to plan for all eventualities. For example, *if* the person transcribing my interviews changes the anticipated deadline for completion, *then* I will (a) transcribe them myself or (b) hire an undergraduate student to assist. Consider how each potential problem would affect your time line.

You must include a budget for a grant proposal. Some funding agencies require the inclusion of the budget in the main body of the document, while others prefer it in an appendix. Wherever you place it, be sure to give the budget the full attention it deserves. Don't estimate the cost of items, but rather check out options and provide quotes. Obtain the latest information regarding standard budgetary items, such as mileage rates, transcription costs, and registration fees. Be sure that the items listed in the budget are eligible for funding. If they are not, seek supplemental funding for those items and provide evidence of that funding. In our experience, funding agencies, particularly for internal or local grants, are often more receptive to proposals that have already established some financial backing.

Oral Presentations

Although this chapter focuses on writing a proposal, we would be remiss if we did not address critical issues related to presenting your proposed study. Indeed, you will commonly present your study to a committee before conducting a thesis or dissertation. Consider the following dos and don'ts when preparing for an oral presentation:

Do the following:

- Prepare a script with content related to all three sections (introduction, literature review, and methods)
- Practice, revise. Practice, revise. Practice, revise!
- Maintain eye contact with your audience (one suggestion to help with this is to have your script in large font on the top 1/3 of every page).
- Construct a slide presentation that highlights key points.
- Provide handouts for committee members.
- Anticipate questions and draft answers ahead of time.
- Be concise when describing the literature.
- Be prepared to make changes. The job of your committee members is to offer suggestions for improving your study.
- Be honest. It is all right to admit that you don't know the answer to a question.
- Be enthusiastic about your study.
- Dress and act professionally.

Do not do the following:

- Act defensive.
- Simply read your slide presentation aloud without any explanation.
- Explain every single study you have read like a verbal string of pearls.
- Get carried away, speaking beyond the given time frame.
- Forget to thank the members of your committee and audience.

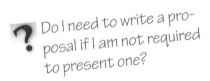 Do I need to write a proposal if I am not required to present one?

Yes! You must complete a detailed proposal even if you are not required to present it. Remember, if you fail to plan, you should plan to fail.

Summary

We cannot stress enough the importance of a well-written proposal. For most of your audience members, it is their first introduction to your study and to you as a writer. As the old adage goes, "You never get a second chance to make a first impression." Plan your study thoroughly and logically. Do your homework and be sure that you can articulate the importance of your study, how it fits into the current knowledge base, and the unique contributions it will make. Remember, however, that a proposal is a plan. If Mr. Murphy comes to visit, a well-written proposal that plans for contingencies with if-then scenarios will send him on his way!

CONTINUING YOUR EDUCATIONAL JOURNEY

 Learning Through Activity

1. Write a 1- to 2-page narrative outlining your relationship to your topic. What experience do you have with this topic? Why are you truly interested in it? Be honest with yourself.

2. Read the article by Goodwin and Compton (2004) in appendix C and underline the key words or terms that you must understand to fully appreciate the study. Did the authors define those terms to assist you?

3. Search your institution and the Internet for funding agencies that may be willing to provide financial assistance for your study. Which do you think would be the most appropriate for your study and why?

 Checking your Knowledge

1. Which of the following is not a component of a research proposal?

   a. introduction

   b. methods

   c. results

   d. literature review

2. When writing a literature review, you should summarize the results of various studies in sequential order, usually chronologically.

   a. true

   b. false

3. If–then statements help researchers to

   a. be clear about what they plan to do in their methods

   b. plan for contingencies

   c. sort out their ideas

   d. present a problem that can be researched

4. The introduction of a study should contain

   a. background information

   b. statement of the problem

   c. a purpose statement

   d. research questions

   e. all of the above

5. How does a research proposal for a grant agency differ from other proposals?

    a. a time line is required

    b. a budget is required

    c. you must have a large research team to conduct the study

    d. a and b

    e. b and c

 ## Thinking About It

1. You have found a small grant for $1000 to conduct your qualitative research; how will you spend the money?

2. While writing your proposal, you designed a time line for completing your study. Imagine possible disruptions to that time line. How might you plan for these contingencies?

 ## Making a Stretch

These readings will help you delve deeper into the realm of qualitative research.

Denscombe, M. (1998). *The good research guide for small scale research projects.* Buckingham, England: Open University Press.

Golden-Biddle, K., & Locke, K. (2007). *Composing qualitative research* (2nd ed.). Thousand Oaks, CA: Sage.

Locke, L.F., Spirduso, W.W., & Silverman, S.J. (2000). *Proposals that work: A guide for planning dissertation and grant proposals* (4th ed.). Thousand Oaks, CA: Sage.

Schwandt, T.A. (1997). *Qualitative inquiry: A dictionary of terms.* Thousand Oaks, CA: Sage.

8

Writing Your Results and Discussing Your Findings

Learning Objectives

Readers will be able to do the following:

1. Describe participants' demographic information and explain the significance of this process.
2. Describe the structure of a results section.
3. Explain the importance of sharing quotes with a reader.
4. Present quotes that support qualitative research findings.
5. Clarify why pseudonyms are used to present qualitative findings.
6. Identify the influence of a target audience and discipline on the written presentation of a results section.

Components of a Report

The term *research report* refers to the outcome of a completed research study. These reports can take many forms, including a thesis, dissertation, published article, or even a committee report. Regardless of type, the following components should be included in every report:

1. Introduction
2. Review of literature
3. Methods
4. Results
5. Discussion
6. Conclusions

Having read the previous chapters, you are probably familiar with the contents of the introduction, literature review, and methods sections. After collecting and analyzing your data, you will be ready to present your results, discuss them, and draw reasonable conclusions.

Chapter 7 outlines the literature review as a separate chapter or section of a proposal. In research reports or journal articles, the literature review is often included as part of an expanded introduction. Some reports require you to share the entire literature review. In these cases, the literature review has its own section that usually appears right after the introduction. Your discipline will determine whether you must include a review of the literature when you submit an article for publication.

When writing a research report, you present your introduction and methods sections in the past tense; that is, in terms of what you have done, not what you will do. Beyond that, the content of these sections is very similar to the proposal. Obviously, you must revise the content to reflect any modifications to your study or procedures. However, since we have already discussed the introduction and methods sections a great deal, this chapter focuses on how to present your results, discussion, and conclusion.

Results Section

We now turn our attention to writing the results section of a study. This section first discusses how to consider your audience and how to describe your participants. Next, it explains how to organize your results section to provide an overview of the structure. Finally, it explains how to present the emergent themes, including how to introduce and share quotes with your readers.

Presenting the results of a qualitative study is perhaps the most difficult part of the research process, but it can also be the most enjoyable. You may find this process challenging because you must present a clear, fair, and unobstructed view of others' realities, perceptions, perspectives, and beliefs. You must treat those who have provided you with data with justice. You must also be fair to yourself, since you have scoured the data and interpreted the findings. Who knows the relationships among the data better than you?

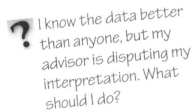
I know the data better than anyone, but my advisor is disputing my interpretation. What should I do?

First, review your findings to see whether you have adequately presented them. Next, ask your participants to verify your interpretations. Finally, gather your evidence and clearly articulate your case to your advisor. If your advisor still disagrees, call a committee meeting and share your data with others.

In writing your results, you have the opportunity to present your position, explain the evidence you have collected to answer your research questions, help others gain tremendous insights about your investigation, and, most importantly, highlight the voices of your participants. Enjoy this culminating process!

Publication Guidelines

The structure of your report may be significantly influenced by your discipline and the journal in which you hope to publish your work. Reports for dissertations and theses are substantially longer than journal articles. Many journals have size restrictions that limit a report's format. Furthermore, universities often have their own expectations about the format of graduate reports.

You may need to present your report using a specific style that suits your format of publication. For example, if you wish to publish an article in the *Journal of Athletic Training*, you must follow American Medical Association (AMA) manuscript style. If you wish to publish in *Research Quarterly for Exercise and Sport,* you must follow American Psychological Association (APA) requirements. Both styles have instructional manuals that will guide your efforts. Be sure to access them in advance so you can structure your report appropriately.

Participant Descriptions

In your study proposal, you articulated criteria for purposefully selecting your participants. Keep this information in your methods section, but enhance it with new, specific data about your participants that you gathered throughout the course of your study. Present as much descriptive information about your participants as possible to help readers determine the transferability of your findings.

Share the participants' descriptive information both generally and specifically. First, present demographic information about your participant pool, such as the number of total participants, their average age, the age range, and a breakdown of participants by gender. Other relevant descriptive information includes titles held (e.g., department chair, health education faculty, and assistant athletic trainer), average years of experience in a particular role, or class standing (senior). Your goal is to carefully craft a description of the group of participants that helps readers understand their background and why they were selected for your study. The following example highlights participant demographic information from a study examining the methods pediatricians use with patients to prevent obesity:

> Pediatricians within the general pediatrics division of the University of Wisconsin provide care to the greater Madison area. Thirty-one spend at least part of their time seeing infants, children and adolescents and therefore were eligible for this study...Together, these physicians have approximately 110,000 patient visits per year...Twenty-four pediatricians (77.4%) were interviewed. They had a mean age of 47 years and had been in practice an average of 17 years at the time of the interview. Forty-two percent were male. Fifty-four percent completed their pediatric residency at the University of Wisconsin and a third practice in private settings. (Gilbert & Fleming, 2006, p. 27)

The authors of this study do a fantastic job situating the context and describing the age, gender, and years of experience of their participants. The authors also provide background information, such as patient visits and geographical location that is relevant to the study. Readers can use this information to discern how the study's findings might apply to their own situations.

In some studies I have noticed that researchers refer to their participants as subjects. Why?

Many journals are so accustomed to using quantitative terminology that they don't consider using the term *participants*. We believe that the term *participants* is more personal and more accurate because qualitative research does not subject people to interventions or procedures.

You should also share specific information about your participants in a table or other systematic format. Researchers commonly use pseudonyms in place of participants' names when presenting demographic information. The specific information linked to pseudonyms helps readers to get to know your participants and relate to their quotes later in the manuscript. You have some flexibility in respect to where you describe your participants. Some journals require you to include this information in the methods section. Other journals view this information as part of your results.

Structuring Your Results

The results section of your study should contain an introductory paragraph, a presentation of the thematic findings supported by quotes, and a conclusion. One way to think of this section is with the old adage, "Let the reader know what you are going to tell them, then tell them what you told them."

The results section of a study may span more than one chapter. Sometimes, especially in dissertations or other large research projects, entire chapters are devoted to individual themes to help readers fully understand the findings. The structure of this chapter reflects the fact that most qualitative publications present the results in a single section. Regardless of how you organize the results section, you must always begin with an introductory paragraph.

Introducing Your Results

The introduction of your results should state both the number of themes that emerged from the findings and their titles. When you use a figure or conceptual model to present your findings, identify it and explain why it provides a good overview. You may need to make a decision about a figure's placement. If it only makes sense after a great deal of evidence is presented, it should probably be presented later in the results section.

Introductory paragraphs also outline the rest of the section so readers know what to expect. Consider the following two examples of how to introduce result sections. The first example is from the *Journal of Sports Rehabilitation*. Pizzari, McBurney, Taylor, and Feller (2002) conducted a qualitative study of rehabilitation programs to identify variables that influenced participants' adherence to the process. At the end of the section on data analysis, they provide a sentence that sets up the first paragraph of the results. We present both the transitional sentence and the paragraph of their results:

Codes were collapsed by grouping together related or similar codes under new headings, and coding was redefined and united until 3 main themes emerged.

Results

The 3 categories of variables identified by participants as influencing rehabilitation adherence were environmental, physical, and psychological factors. Figure 1 shows a flowchart derived from thematic coding of the variables and their thematic groupings. (p. 94)

Here the authors identify the three emergent themes and direct the reader to a figure that illustrates their results. Although the authors do not explicitly state what they plan to share with readers, the remaining text uses subheading to highlight themes so readers can easily understand the content's organization.

The second example is from the journal *Women and Health*. Ball, Salmon, Giles-Corti, and Crawford (2006) studied physical activity of women from different socioeconomic backgrounds. Here is how they introduced their results:

Eleven main themes were identified. These were: participation in different types of physical activity, physical activity history, lack of time, planning/routines, lack of motivation, value of sedentary behaviours, social constraints/support, the work environment, local neighbourhood safety and aesthetics, local physical activity facilities, and financial costs of physical activity. The themes are described below. (pp. 98-99)

In this example, the authors clearly articulate the number of themes that emerged from the study. They also explicitly state that they plan to describe each theme. The introduction to your results should lead to the presentation of each theme, which is the most critical aspect of the results section.

Themes

Once you have introduced the findings, you need to present the emergent themes. Use subheadings to organize your content and present the information in a logical manner. You should present and discuss the themes in the order that they appeared in the introductory paragraph.

As you present the emergent themes, remember that simply paraphrasing what you have learned and showing a graphic do not convey a depth of understanding and insight. Instead, you must use participants' quotes to support your emergent themes and build a case of truth and believability for your reader.

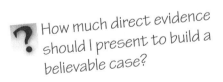 How much direct evidence should I present to build a believable case?

The quantity of direct quotes depends on your data. You need to find a balance that works for you between the voices of the participants and the researcher.

Researchers commonly present a theme by explaining the meaning of the thematic title and its key aspects. They support the explanation with a direct quote from at least one participant (assuming that they have conducted interviews). For example, in their 2002 article, Pizzari et al. presented the theme of environmental factors in the following manner:

The major environmental factor influencing the completion of home exercise was reported to be lack of time. An abundance of work, holiday, family, and social commitments depleted the amount of time available for rehabilitation. With "just too much (going on) in life" (Belinda), some found that "trying to do rehab around that did get quite difficult" (Mary). (p. 94)

The authors first identified lack of time as a key environmental factor. They then provided short quotes from two participants as evidence.

Another important facet of the preceding example is the authors' use of pseudonyms after the participants' quotes. Readers now know that the quotes came from two different sources. By sharing quotes from various participants throughout the results section, researchers can give readers a sense of data saturation. If they fail to share pseudonyms in the research report, readers will have no way of knowing whether quotes come from one source, two sources, or almost all of the participants.

In presenting the emergent themes of a study, researchers must not only explain the findings from their perspective, but also clarify the themes by sharing direct quotes from participants. Both practices highlight important issues related to the emphasis researchers place on participants' voices and the way that they present quotes. We begin by discussing the balance between the presentation of researcher and participant voices.

Dominant Voice   When explaining your study's emergent themes and findings, you must decide which participant quotes to use, how many quotes to use, and how much of your voice to share in the results section. These important considerations all relate to the issue of voice emphasis. In other words, which perspective do you wish to highlight more, the participants' or your own?

We believe that you must strike a balance between the two voices to present the results in a meaningful, compelling, and accurate manner. Having served as the instrument for both data collection and analysis, the researcher (or research team) has listened to participants during interviews, watched their actions, and, in many instances, has reviewed numerous related documents. Moreover, the researcher has filtered out unnecessary information, considered important information from many different viewpoints, and crosschecked a great deal of evidence. Who could better explain the emergent findings than the researcher? However, the process of qualitative research is geared toward understanding participants' perceptions of their experiences. The voices of participants highlight the significance of concepts and issues within a given context. From that perspective, who is better equipped to provide a depth of understanding than the participants themselves?

We hope that the preceding paragraph builds a convincing argument for the validity of both voices. Figure 8.1 graphically displays the range of options for balance of voices in the results section. Examples follow the discussion of each option. Please note that we do not rank any single method above another. Each method has advantages and disadvantages.

- *Quadrant 1: High presentation of researcher's voice; low presentation of participants' voices.* In this method, the researcher's voice is dominant in the explanation of the findings. Subsections of results contain few, if any, quotes from participants, and often simply explain the findings from the researcher's perspective. This method is necessary for researchers who consistently hear comments from participants that, however meaningful to the research questions, are extremely pithy.

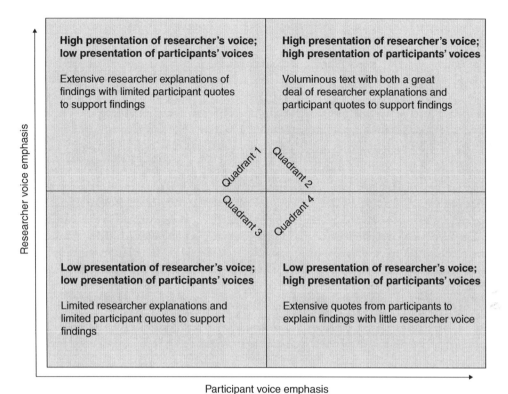

Figure 8.1 Options for voice emphasis.

In this example from his 2002 study, Pitney presented data related to an emergent theme without participant quotes:

> Aside from facilitating learning, networks were also used to provide help and social support (concepts B and D, category 2). For example, Reginald explained that he received a call from a high school [athletic trainer] in a nearby suburb. The [athletic trainer] experienced a death of an athlete approximately 6 months earlier and called to get contact information for another area [athletic trainer] who experienced an athlete's death more recently because he wanted to offer his support. (p. 290)

In spite of its lack of quotes, the excerpt fully describes the situation and offers information that supports the emergent category. Researchers may find this method of presenting data useful, but they should not rely on it for entire sections of a report. It is necessary to share participant quotes to enhance the believability of the findings.

• *Quadrant 2: High presentation of researcher's voice; high presentation of participants' voices.* Some studies have many emergent categories and subcategories. When thorough explanations from participants provide rich details about a finding, researchers may need to fully explain the emergent categories and subcategories before sharing lengthy quotes that support the claims. A study from Pizzari et al. (2002) illustrates this presentation style. In this excerpt, the authors explain why the type of support a physiotherapist provided to patients was critical to their adherence to the rehabilitation program:

> The most significant part of the rehabilitation process for most participants was their interaction with their physiotherapists. Physiotherapists were described as friendly, knowledgeable, and supportive, and most respondents indicated that their positive relationship with the therapist helped with attending the clinic and completing rehabilitation appointments.

> The informational and emotional support provided by physiotherapists throughout rehabilitation was important to all participants. Particularly in the initial stages of rehabilitation, information regarding the injury and rehabilitation process was thought to be vital for adherence. When information was lacking, nonadherence resulted.

> "I started physio 3 weeks after my operation...The people at the hospital didn't really inform me of what I had to do. I mean maybe it was naïve to think I'd get a phone call to say 'You have to start physio,' but I suppose that's what I was thinking at the time. I wish I'd have known and I would have started it earlier. I mean, I knew I had that sheet from the hospital, on 2 separate occasions, and I wish they'd have stressed more that the first couple of weeks was the most important, just to keep it moving, because I don't think I moved it enough. And I think that...took me longer to get started. So, the first couple of weeks of physio was sort of like behind" (Belinda). (pp. 96-97)

As you can see, the authors begin with their own interpretation of the findings, and then support it with a long participant quote. This presentation style offers a comprehensive picture and fully describes the findings.

• *Quadrant 3: Low presentation of researcher's voice; low presentation of participants' voices.* In some instances, the presentation of qualitative findings highlights neither the researcher's voice nor the participants' voices. This situation is rare, but writers are sometimes forced to use this presentation style by a journal's space limitations. Consider the following example from a study about participants with multiple sclerosis who completed a resistance-exercise program (Dodd, Taylor, Denisenko, & Prasad, 2006). The authors presented data within the theme of "positive physical outcomes." Here is the introduction of the theme:

> Not unexpectedly, all nine participants reported that they felt stronger at the end of the program. Four participants noted improved endurance and two participants noted improved flexibility: participant 1 "noticed more dexterity in my fingers," while participant 6 observed "less stiffness...I always used to feel stiff."

> Five participants also reported that various activities of daily living had improved. Participant 2 stated that with walking "I used to stagger, trip...and since I've been doing strengthening I've noticed it a lot less." Participant 6 felt that "walking up a set of stairs with shopping doesn't seem to concern me too much anymore." (p. 1130)

As you can see, the authors neither provide lengthy quotes from participants, nor offer an exhaustive interpretation. However, the authors also explain their findings with a visual diagram. We would like to note that this style of emphasis works within the context of this particular study. The authors did not need elaborate explanations to convey their findings.

•   *Quadrant 4: Low presentation of researcher's voice; high presentation of participants' voices.* Some authors prefer to stay out of their presentation of the findings. Instead, they use extremely long quotes from participants in the results section and offer very little interpretation. Researchers commonly use this approach in narrative studies, which focuses on the life stories.

This presentation style is also rare, and is not compatible with the space limitations of journal publications. Theses and dissertations have fewer space requirements, so authors have more freedom to use lengthy participant quotes. Consider the following example that emphasizes participants' voices:

Emotional Support

According to many of those interviewed, emotional support was the number one type of support needed by an injured athlete. One coach said, "The greatest need of an injured athlete is emotional support….making sure that emotions are getting taken care of." Another coach advised, "…be consistent with injury as with other psychological issues with athletes." An athletic trainer agreed by saying, "The mind is very powerful and controls everything in the body….it is important to support them emotionally in order to keep their mind focused on the task of rehab." Another athletic trainer said, "It is all about emotions when injured." One athlete said, "Emotional support is most important because when injured emotions are going all over the place….you need people to keep you going."

Another form of emotional support is that of encouragement. One athlete believed, "….encouragement, that's what it is all about." Other injured athletes agreed by saying, "Knowing people are behind you….encouragement and reinforcement are great," "Real positive encouragement is needed and helped me…see the brighter side of things," and "Encouragement from coach really helped me, emotionally and all."

Coaches also realized the need for encouragement saying, "…me supporting them and encouraging them is very helpful emotionally," and "…encourage the injured athlete…it is hard work to heal and get back to playing." One athletic trainer said,

"My job has a lot of different responsibilities….I need to work with the body to heal and recover, but it is also my job to encourage them [injured athletes], encourage them in healing and encourage them in their mindset…Not all athletes feel the same way about injury, but [it] is important to make sure they feel encouraged and are emotionally stable….there are many emotions present when an athlete gets injured, part of my job is to deal with those."

Yet another important component of emotional support is understanding. Athletes commented, "The athletic trainer had an understanding of the situation," and "Coach understands and works with me." One athletic trainer also recognized the importance of the coach understanding in stating, "It is important for the athlete to have understanding from the coach." When talking about support given to injured athletes, one coach said, "The coach and athletic trainer need to respond well to injury and understand the athlete might have a wide variety of feelings at first." (Borseth, 2004, p. 47)

Consider these four styles when presenting your data. Hold your readers' interest by varying your emphasis of voices. We suggest that you borrow techniques from each quadrant as you present your results. The next section provides suggestions about presenting your quotes.

Quotes as Evidence   In presenting your themes, you must share participant quotes. The quotes are the "raw data" that informed your coded concepts in the first place which were then used to form your themes. Use them to bring your results section to life and help readers make clear links between the quotes and the themes.

Quotes can be presented in many ways. You can use long, block quotes; quotes of medium length that are integrated directly into a sentence; or very short, pithy quotes. This section explains each type of quote. As you read it, keep your audience in mind. Remember that your readers will appreciate variety of presentation. Therefore, we recommend using quotes of different lengths from various participants. Compare this practice to sentence structure. If every sentence in a paper were complex, readers would soon lose interest. You also need to use simple sentences. A combination of simple and complex sentences creates a unique rhythm and makes the text easy to read. Participants' quotes can achieve the same balance.

- *Use block quotes.* Long participant quotes give readers a rich sense of context, and almost allow them into the minds of participants. Consider Pizzari and colleagues' 2002 explanation of how time restraints affected their adherence to rehabilitation:

> Being a mother…you have to prioritize and you get to be good at doing that you know, so you sort of like think "I've got to do this, I've got to do that," and in your mind you're already thinking "I've planned this and planned that," and we can do it you know…you just prioritize what you need to do and get the job done (Jane). (p. 95)

- *Integrate quotes into a sentence.* Sometimes a short, concise quote adequately conveys meaning and offers evidence for a claim. Short quotes can be worked right into the body of a sentence. For example, in a study on disability and the mediating effects of physical activity, Goodwin, Thurmeier, and Gustafson (2004) presented a theme titled "Don't treat me differently," and explained how one participant described a normalizing experience. Here is how the authors integrated a quote into the body of a sentence:

> Although not all of the participants drove cars, those who did spoke of what a normalizing experience it was. Fellow drivers were unaware of the participants' impairments thereby eliminating the dependency assumption. Liz recollected, "I'm hollered and honked at just as much as anyone; I get the same treatment. I love driving because when I'm driving I'm just me and nobody has any reason to believe I'm not like any other driver." Meg commented on how impressed her classmates were. (p. 388)

- *Use pithy quotes.* Sometimes, no matter how many ways you ask a question, participants simply do not give an elaborate response. Qualitative researchers who primarily interview teenagers often encounter this problem. Don't fret if your evidence mostly consists of very short comments. You can still use these quotes. In fact, the presentation style from quadrant 3 is very appropriate for short, pithy quotes.

Remember that you can only work with the data you have obtained, so you must do the best you can.

We refer back to the work of Dodd et al. (2006) because they did a good job of using and presenting pithy quotes. Another theme they presented was related to the positive social outcomes patients with multiple sclerosis experienced by participating in resistance-training programs. The authors weave and integrate quotes from several different participants into their sentences:

> The eight participants who completed the programme valued the social aspect of the programme, including the companionship and friendships that developed. Participant 2 stated that she had "formed new friendships," participant 8 "loved the atmosphere...the group was nice," and participant 6 "enjoyed everybody's company...the camaraderie." (Dodd et al., p. 1131)

You have many options when presenting your evidence. Remember to vary your use of quotes, as well as the length of your own commentary. From a reader's standpoint, the results will flow much better and make the reading process more enjoyable.

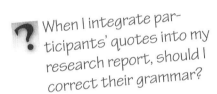

When I integrate participants' quotes into my research report, should I correct their grammar?

You must maintain the integrity of the meaning that the participant provides and be sensitive to their verbal and cultural expressions. However, you must also ensure that readers will understand the quotes and their connection to the theme. Occasionally, you may need to remove extraneous utterances like *um* or add a clarifying term or phrase in brackets.

Summary of Results

Once you have presented your emergent themes, you must summarize them. When a results section has many themes, lots of quotes, and rich information, you should provide a paraphrased summary. This section gives you one last chance to concisely state the thematic findings. However, summaries are not required. Many published studies omit this part of the results section because of space limitations. Another reason a summary may be omitted is that the discussion section of a manuscript often restates the study's purpose and findings.

Discussion Section

In a research report, the discussion should present a thoughtful discourse that links your findings with relevant literature. It should follow a coherent introductory paragraph with a systematic presentation of findings that are situated within the context of the literature. The discussion concludes by presenting any implications the findings present for theory, practice, or policy.

Introducing the Discussion

The introductory paragraph should restate the purpose of the study, and then provide a general statement about its findings that reminds readers of the context. Although this practice may seem a little redundant, remember that some readers skip straight to the discussion section to discern the essence of a study. When introducing your discussion, remember to revisit your purpose statement and research questions. If you fail to do this, the readers may feel a bit cheated. Consider this example from a study from Hodge, Tannehill, and Kluge (2003), who explored practicum experiences for physical education students:

> The purpose of this phenomenological qualitative study was to explore, using self-reflective journaling, the meaning of practicum experiences for PETE students enrolled in an introductory [adapted physical activity course] with an inclusion-based practicum requirement. Guided by weekly probes from the course instructor, these students reflected on and shared their thoughts, feelings, impressions, beliefs, and attitudes about working with youngsters with and without disability in physical activity settings...These self-reflections impelled them to identify issues, to address problems, and to think critically about how best to enhance the physical activity setting for youngsters with and without disabilities. (p. 382)

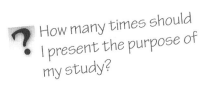
How many times should I present the purpose of my study?

In an ideal world, all readers would view our work in the order we wrote it. However, just as you may have jumped around while reading this book, students may elect to read your work in the order that makes sense to them. Therefore, you should reiterate the purpose statement at the beginning of a discussion. In a thesis or dissertation, expect to restate the purpose statement at the beginning of each chapter. Don't fight it, just do it!

In the preceding example, the authors restated their purpose before discussing the primary form of data collection. They next articulated how this form of data collection led them to their overall findings. From that point, the authors were ready to systematically explain each of the key findings and its relationship to the literature base.

Body of Text

When writing the body of the discussion, be sure to systematically discuss your findings in the same order that you presented them in the results section. This practice adds logic and order to your overall presentation. When discussing your findings, you have some freedom to speculate about the meaning of the findings, but readers will expect you to compare and contrast your findings with existing literature.

You have two options for relating your study to existing literature. You can first discuss the finding itself, then explain the relationship. Your other option is to discuss the literature before articulating your findings. We present examples of each technique. The first example presents the finding from the study first. The second example leads with a literature citation. Both methods of presentation are appropriate.

 I know I need to link my results to the literature in the discussion section. Can I introduce new literature at this point, or should I only refer to literature included in the review?

Again, it depends on your intent! Are you introducing new literature because your findings led you down a different path, or because you simply missed these articles in your literature review? The former might be anticipated, but the latter is unacceptable.

In a study of women with disabilities and their perceptions about aging, Goodwin and Compton (2004) relate the loss of physical freedom, one of the findings from their study, with two sources from the literature:

> The potential for decreased independence with age elicited strong psychological feelings about the women's quality of life. The women were disheartened about the prospect of requiring the assistance of others. A depressed mood state has the potential to negatively impact the incentive needed for the women to continue to be causal agents in their own health and physical well-being and maintain physical function as long as possible (Dunn, Trivedi, & O'Neal, 2001; Moore et al., 1999). (p. 135)

In a study of the professional socialization of athletic trainers who work in the intercollegiate setting, Pitney (2006) related findings in the literature with those in his study:

> Bureaucratic influences have the propensity to create routinization of work and managerial control, increase volume of work, and down-grade job-related tasks and skill levels. The participants experienced several of these aspects, including increased work volume, impersonality (lack of support and appreciation by administration), and hierarchy of authority. (p. 192)

Results that are noteworthy and very unique require discussion, regardless of whether comparable literature exists. Hodge et al. (2003) used this tactic in discussing a noteworthy finding:

> In addition to the teaching variables listed above, and not surprising, journal entries included reference to organization and class management. They reflected on the importance of establishing rules and routines as helping to manage the youngsters' behaviors. Moreover, students pointed out that social reinforcement (e.g., verbal praise), token economy system, and physical activity reinforcers following Premack principles were useful strategies for managing youngsters' behaviors. Further, they reflected on the humanistic, educational, and social values of inclusive physical activity programming. (p. 397)

Even though the preceding paragraph does not articulate a relationship to other sources, the authors discuss informative findings that may help readers, particularly educators, understand how the journaling process identifies helpful strategies for class management.

Limitations

No study is perfect! Acknowledge your study's limitations. In other words, every study has at least one drawback. You have an obligation to share them with readers. For example, one key limitation of qualitative research is that the results cannot be generalized for a vast population. When you write a research report, consider placing your limitations at the end of the discussion section, before you conclude your study. This practice is different from a proposal, which shares potential limitations in its introduction.

Conclusions and Implications

End your report with conclusions, implications, or a combination of both. The conclusion should provide answers to your research questions and summarize the study's findings. You can also identify practical implications that your study generated. Try to add a new level of significance for your findings (Booth, Colomb, & Williams, 2003), and point out practical solutions for any identified problems.

The conclusion section commonly includes suggestions for future research. For example, Hodge et al. (2003) wrote, "To learn more about how to guide students in the journaling process, future research could focus on the experience of journaling as a form of self-reflection" (p. 397). This is an excellent place for novice researchers to obtain ideas for their own projects.

Final Review

Once you have completed your results and discussion sections, you should revisit your title and modify it, if necessary. Many journals prefer a title to capture the study's conclusions. Remember that the title of your study will help others locate your study. It should succinctly capture the content of your study, but be general enough that readers will be able to access it with a research database or search engine.

You must also create an abstract that provides a concise (usually no more than 400 words) summary of your study. The abstract should include information related to the study's objective, design, participants, methods of data collection and analysis, results, and conclusions. Most journals also require you to include key words at the end of your abstract. The key words should differ from those in the article's title.

 Does the title really matter?

Yes! The title is important because it is the first part of the study your readers will see. You must strike a balance between catchiness and content. The title must pique the readers' interest but be easy to locate with current research databases. Conduct a search using

> common descriptive terms for literature on your topic. Ask yourself whether you would be able to find your own study if you were to redo your literature review.

As you assemble your manuscript, visit the author's guide for information about the order of the materials, and how to display various items. For example, you may be required to present all of your figures at the end of the manuscript, rather than within the body of text. Every journal operates differently, so be aware of the specific guidelines.

Summary

The results section of a well-written qualitative study uses quotes to support emergent themes. As authors interpret the findings and paraphrase the data, they must provide participant quotes that support the claims and substantiate the findings. They should share quotes in a variety of ways to make the report easier to read.

In the discussion section, authors should first remind readers of the study's purpose, then compare and contrast the findings with related literature. They should draw reasonable conclusions and identify implications for future research based on the study's results.

CONTINUING YOUR EDUCATIONAL JOURNEY

 Learning Through Activity

1. Examine the results section of the article by Pope and O'Sullivan (2003) in appendix B for descriptions of participants. How many people participated in the study? What were their ages, genders, and ethnicities?

2. Evaluate the authors' use of participant quotes. In which quadrant did they present the majority of evidence?

3. Read the entire article and write a 400-word abstract. Please challenge yourself and do not use aspects of the abstract provided!

4. Identify the limitations of the study described by the authors. What other limitations can you identify?

Checking Your Knowledge

1. The structure your report takes may be significantly influenced by your _____ and the _____ in which you hope to publish.

   a. research budget; Web site

   b. discipline; journal

   c. writing ability; Web site

   d. research budget; journal

2. It is critical to describe your participants, but you must also protect their identities. Which of the following is an acceptable way to protect the identities of research participants?

   a. Never use their quotes.

   b. Always provide their real names.

   c. Use pseudonyms.

   d. Consider using only sources that allow you to publish their names.

   e. b and c

3. How can you use quotes to support your research findings?

   a. Use block quotes.

   b. Work your quotes into the structure of the sentences.

   c. Present the data by paraphrasing its meaning.

   d. All of the above are useful ways of supporting your research findings.

4. When writing a discussion section, it is not important to utilize existing literature.

   a. true

   b. false

5. Which of the following is true of the discussion section of a research report?

   a. You should remind the reader of your study's purpose.

   b. You should compare your research findings with those in existing literature.

   c. You should contrast your research findings with those in existing literature.

   d. b and c

   e. a and b

 Thinking About It

1. In your opinion, what is the most important aspect of a discussion section?

2. Why should authors provide practical suggestions about the implications of a study?

3. What ethical obligations do researchers have while presenting the findings of a study?

4. When presenting quotes, some researchers use pseudonyms and others use participant numbers in place of names. Identify the advantages of each method.

5. One of your participants is not a native speaker of English. The transcript of the conversation is very fragmented. How might this affect your presentation of the results?

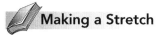 **Making a Stretch**

These readings will expand your scholarly communication skills.

Knight, K.L., & Ingersoll, C.D. (1996). Optimizing scholarly communications: 30 tips for writing clearly. *Journal of Athletic Training, 31,* 209-213.

Knight, K.L., & Ingersoll, C.D. (1996). Structure of a scholarly manuscript: 66 tips for what goes where. *Journal of Athletic Training, 31,* 201-206.

Continuing Your Qualitative Research Journey

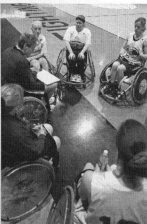

Part IV begins with chapter 9, which presents specific forms of qualitative research. It focuses primarily on grounded theory, ethnography, and phenomenology, but also presents qualitative case studies, narrative inquiry, and action research.

Chapter 10 addresses how to evaluate qualitative research. It provides a variety of questions to consider as you become a critical consumer of qualitative research.

Our final chapter examines the most common arguments and assumptions made against qualitative research and offers suggestions on how to respond to critics. It also explains how and when to combine qualitative research with quantitative research. It concludes with practical advice and resources for continuing your journey as a qualitative researcher.

Guiding Questions

Consider the following questions before reading part IV. They will guide your examination of each chapter.

1. What forms of qualitative research exist? How are they similar to and different from one another?
2. What should you know when evaluating a qualitative research study?
3. How does your purpose for evaluating a study influence the questions that you ask?
4. What are common criticisms of qualitative research?
5. When is it appropriate to design a study that includes both qualitative and quantitative methods?

The following figure illustrates the content, connections, and organizational configuration of part IV.

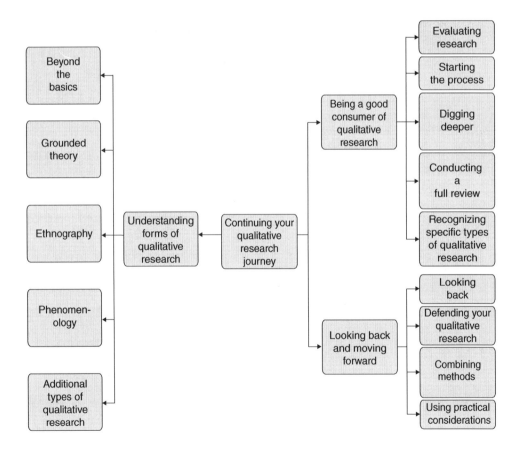

9

Understanding Forms of Qualitative Research

Learning Objectives

Readers will be able to do the following:

1. Identify various forms of qualitative research.
2. Explain the focus of grounded theory, phenomenology, and ethnographic studies.
3. Compare and contrast methods of data collection and analysis used with grounded theory, phenomenology, and ethnographic studies.
4. Describe research products associated with the forms of qualitative inquiry.

Beyond the Basics

The preceding chapters introduced the basic, generic approach to qualitative research. We borrow the term *generic* from Merriam, (1998) who articulates that this is indeed the most popular way to conduct qualitative research. The premise of our text was that the way in which data are collected and analyzed is similar regardless of the form of qualitative study that is performed. Thus far you have learned the basics of inductive analysis that is transferable to all forms of qualitative inquiry. A basic approach to qualitative research is a very pragmatic way to approach the research process.

However, if you want to learn more about qualitative research and to conduct future studies, you should know that many forms of qualitative research exist. In order to be an educated consumer and astute researcher, you must be able to recognize the various forms of inquiry and understand the focus, method, and product of each.

The differences between the forms of qualitative inquiry are quite interesting. We believe Schram (2006) said it best with the following quote: "[w]ays of looking—observing, asking, and examining what others have done—are notably similar across many qualitative research traditions. Ways of seeing—encompassing underlying intent, guiding concerns, focus, and perspective—are not so similar" (p. 94).

As we explore the common forms of qualitative research, keep in mind that no one way of examining the social world of others is most correct (Holloway, 2005). Indeed, there is a great deal to consider when choosing an approach. The form of qualitative research that you choose depends not only on the purpose of your study and research questions, but also on your philosophical stance and personality (Holloway, 2005). This chapter, therefore, introduces various forms of qualitative research, including grounded theory, ethnography, and phenomenology. It discusses the focus, method, and product of each form, and also lists examples of research purposes and questions for each. The chapter concludes with a small section on forms that are less common, including qualitative case studies, action research, and narrative inquiry.

Grounded Theory

Grounded theory is an approach to qualitative research that originated in the Chicago School of Sociology between the 1920s and 1950s (Kendall, 1999). Two American sociology researchers, Glaser and Strauss, are credited with bringing direct attention to this method of inquiry. In the 1960s, they created a very systematic way to analyze qualitative data and generate theory (Bluff, 2005).

The initial process for developing grounded theory was termed the *constant comparison method* (Glaser & Strauss, 1967). It is composed of the following distinct stages:

1. Identifying and comparing information or incidents
2. Developing categories and constructing subcategories
3. Delimiting the theory
4. Explaining the theory

These stages, although still used in many qualitative studies today, have been refined. In fact, grounded theory has evolved considerably since the 1960s, largely due to the work of Glaser and Strauss (Hallberg, 2006). However, many authors still refer to grounded theory as the constant comparative method. An article by Mensch and Ennis (2002), for example, utilizes this earlier approach, while the article by Pitney (2002) references a more contemporary method (see appendix A). Regardless of their differences, the research focus is the same in both examples.

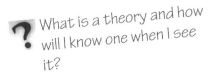 What is a theory and how will I know one when I see it?

Good question! A theory is a set of statements, hypotheses, or propositions that explain a particular phenomenon or process (Ary, Cheser Jacobs, & Razavieh, 2002). Theories come in many forms. Those that are more formal are associated with a label or a researcher's name. For example, Bandura's social learning theory is often referred to as Social Learning Theory. Less-formal theories are displayed as simple statements or hypotheses.

Focus

Grounded theory seeks to explain what is occurring in a social context and aims to develop a theory (Annells, 1996). Therefore, it focuses on social processes. The example of the study by Pitney and Ehlers (2004) focuses on process by examining the way in which athletic training students are mentored. Moreover, the study examined in chapter 2 focuses on how athletic trainers were socialized into their roles in high schools.

Focusing on process generates research questions such as, "How are students mentored to conduct research during undergraduate kinesiology programs?" or "How do exercise physiologists formulate their perceptions of obesity?" A rigorous form of inquiry must be used to generate a theory.

Methods

Grounded theory uses many forms of data, but participant interviews are most common. Alternate but very viable methods are the use of documents (Pandit, 1996) and observations as primary data sources.

The steps for data analysis are very prescriptive and the stages of analysis are clearly identifiable (Glaser, 1978; Strauss & Corbin, 1990). In grounded theory, the researchers concentrate first on coding the available data as they are collected. Coding is the process of identifying and labeling pieces of information, concepts, or experiences that pertain to the research questions. The first stage of this procedure is called *open coding* (Strauss & Corbin, 1990). Open coding involves creating and organizing categories, and is similar to the thematizing process identified in chapter 4.

The categories that emerge from the data during open coding are "an essential aspect of transforming raw data into theoretical constructions of social processes" (Kendall, 1999, p. 745). The process of open coding also identifies subcategories. Constant comparison is an important aspect of this process (Glaser & Strauss, 1967). As the name suggests, researchers continuously compare data codes for similar information in order to create distinct categories.

Strauss and Corbin (1990) identified axial coding as the second stage of analysis. Axial coding is the process of making connections between the categories and subcategories. Researchers must examine several components related to causes, contexts, contingencies, consequences, covariances, and conditions (Glaser, 1978, as cited in Bluff, 2005). Researchers ask the following questions to discover these emerging relationships:

- Do the instances in one category seem to cause or create instances from another category?
- Does the context of the study influence the events or processes that have been discovered? If so, how?
- Are there consequences when an event occurs? If so, what are they?
- Are there contingences that occur only when another event unfolds?
- When one situation changes, does another change occur?
- Under what conditions do experiences and processes occur?

By asking these questions, researchers identify the relationships between the emergent categories. This process of coding helps researchers conceptualize and explain what is happening with the study's participants.

The selective coding process is the last phase of analysis. In selective coding, researchers identify a core category to which all other categories are related. They develop a set of explanatory concepts when they describe the relationship between other categories and the central category. Thus, the theory emerges (Pitney & Parker, 2002). The creation of theory requires a great deal of thought, logic, and creativity from the researchers as they fully examine relationships in the data.

Products

The name *grounded theory* foreshadows its product—a theory, or set of explanatory concepts, is produced. The product of grounded theory is an understanding of what is happening with an issue under investigation.

The theory must be based, or grounded, in data obtained during the investigation. Although information from previous studies may inform the process, researchers must rely on data from interviews and observations to develop a theoretical explanation of the process which they studied.

Many researchers choose to portray their theories as a model. Models or diagrams can give a visual picture of what has occurred and how events relate to one another.

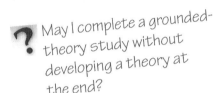 May I complete a grounded-theory study without developing a theory at the end?

No! If you follow a specific method, you should hold true to its tenets. You have not finished a grounded-theory study until you have formulated a theory based on the data you collected.

Purpose and Questions

Here is an example of a purpose statement that would direct you toward a grounded-theory study: "The purpose of this...investigation is to understand the process by which clinical decision making is learned by athletic training students" (Pitney & Parker, 2002, p. S171). A subsequent research question might be: "In what way do interactions with coaches influence the clinical decision making of athletic training students?" (Pitney & Parker, 2002, p. S171). Appendix A contains an example of a grounded-theory study.

Ethnography

Because ethnography is rooted in anthropology, it may conjure up images of exploring the culture of remote people in a secluded land. From a more contemporary perspective, researchers may find cultures and subcultures everywhere: in hospitals, outpatient rehabilitation clinics, physical education classes, or athletic-training education programs.

Focus

Ethnographic qualitative research involves describing and interpreting a group's culture (Lancy, 1993). Cultural aspects of interest to ethnographers include a group's beliefs, values, and knowledge. Culture tends to be "coded in symbols, the meaning of which have to be learned, be it language or [behavior]" (Sharkey & Larson, 2005, p. 169).

To clarify the focus further, ethnographic studies attempt to understand how groups create and negotiate meaning, how relationships develop between people, or even how people create and institute policy (Beach, 2005). Additionally, ethnographers hope to identify the structure of a culture, such as whether a hierarchy exists, what the punishments and rewards are like, and whether there are any rituals. Pope and O'Sullivan (2003) conducted an ethnography of sport experiences at a large urban high school. The authors identified and described a student hierarchy and provided insights about the nature of student relationships and roles within the structure of the school.

To learn aspects of a culture, researchers must become part of the group they are studying (Sharkey & Larson, 2005). You cannot learn a culture from the outside looking in, you must examine it from the inside looking around. Culture shapes who we are and how we think. Therefore, when researchers conduct an ethnography, they attempt to truly gain an insider's perspective by learning norms and attempting to understand the "…social world of people." (Sharkey & Larson, 2005, p. 169).

Methods

Many ethnographers will use all types of data, including quantitative data. For example, Pope and O'Sullivan (2003) fully described the high school's size, ethnic profile, and even the attendance at recreation centers within the local community. Combining census data with qualitative data allowed the authors to provide a rich description of the setting.

Fieldwork is the cornerstone of the ethnographic research method, and includes going to the cultural setting to collect data. Ethnographers visit settings and stay for prolonged periods of time to observe and interview. Due to the extensive period of enculturation, researchers may have difficulty gaining access to the environment. The ethnographer, therefore, must be adept at negotiation.

Most ethnographers place more emphasis on observation as a method for collecting qualitative data. Although it is important to learn about participants' beliefs and values by interviewing them, ethnographers want to observe these qualities in a natural setting. The forms of observations discussed in chapter 4 are relevant to ethnographic studies. Spradley (1980), a leading ethnographer, suggests that researchers begin observation by simply describing the physical setting and participants. Descriptive observations will soon lead to focused observations in which researchers deliberately identify what they need in order to answer the research question. Finally, the observation becomes more selective in nature as the researchers identify activities, events, and specific locations for future observations that will address the study's purpose. Think of this process as gaining a sense of the whole before looking at a culture's parts.

Researchers should conduct an open, broad observation before focusing attention on specific aspects to attain the openness of observation associated with ethnography (Baszanger & Dodier, 1997). In other words, researchers should not bring preconceived notions to the task of collecting data. Instead, they must be open to true discovery about how people interact with others in the group (Baszanger & Dodier, 1997).

Another key aspect of the ethnographic method is the use of field notes. Field notes involve logging or documenting what is observed and learned about the cultural setting. Ethnographers also note what their observations mean in the broader cultural context. Bogden and Biklen (2007) describe field notes as "the written account of what the researcher hears, sees, experiences, and thinks in the course of collecting and reflecting on the data in a qualitative study" (p. 119).

The structure of field notes varies from researcher to researcher, but all must record the following two components:

1. A description of what has occurred
2. What this event means for the broader research purpose

Of all the many forms of qualitative research, ethnography is perhaps the least structured. The quality of being both systematic and flexible is exemplified by ethnography because researchers explore a culture which they do not know. Ethnography has been described as "…an unstructured approach to research where the researcher needs to be explorative and flexible to 'follow the data,' making decisions throughout the research process about what, where, and when data will be collected" (Sharkey & Larson, 2005, p. 169). As personal relationships are discovered and social interactions unfold, the researchers may need more time to understand important facets of the findings.

 I understand that long-term fieldwork is important, but is it possible to wear out my welcome at the research site?

You must remain culturally entrenched until you have gained the insight necessary to thoroughly explain and describe the culture. In this case, time is measured in understanding rather than in hours.

Products

Every form of research has an outcome. In ethnography, the outcome presented in an article is a rich description of a specific culture and the social dynamics that occur. The written presentation of an ethnography is largely in narrative form. As a reader, you should expect rich explanations of the culture or subculture that include information related to the physical environment, interpersonal relationships, hierarchy, rituals, norms, and values of a group.

Purpose and Questions

Here is a brief example of a purpose statement that would lead a researcher toward an ethnographic study: "The purpose of this…investigation is to gain insight and understanding about the culture of learning as it relates to developing clinical decision making skills" (Pitney & Parker, 2002, p. S171). A subsequent research question might be, "What are the norms and values that guide decision making in this particular clinical culture?" (Pitney & Parker, 2002, p. S171). Appendix B contains an example of an ethnography.

Phenomenology

Phenomenology is both a research method and philosophy. The philosophical base draws from the pioneering works of Merleau-Ponty, Husserl, and Heidegger (Todres, 2005).

The philosophical tenets are well beyond the scope of this text, but the final section of this chapter recommends additional sources.

Focus

The focus of phenomenology is how people experience and draw meaning from their worlds. As the term implies, researchers are interested in how participants experience a specific phenomenon. A phenomenological research study focuses on the meaning that several participants assign to a particular experience (Creswell, 2007). Therefore, phenomenology focuses on describing the essence of experience.

The term *essence* means that researchers search for the fundamental nature, real meaning, or core aspects of the phenomenon. The process distills a set of experiences until all that is left is what really matters most, which helps describe a participant's life, world, and perception of a phenomenon.

Phenomena are plentiful in our daily lives. Learning, loving, caring, and hoping are all examples. In athletic training, researchers could frame *loss of function* and *disability* as phenomena and explore the experiences of living with a disability from the perspective of an injured athlete. Goodwin and Compton (2004) conducted a phenomenological investigation to understand the aging experiences of young women with physical disabilities (see appendix C).

Methods

The methods of phenomenology have been described by well-respected researchers, including Giorgi (1985) and Colaizzi (1978). Giorgi explained the phenomenological method as having four steps, while Colaizzi identified 7 steps. Lemon and Taylor (1997) also explained the phenomenological method. We have combined their work to capture the method with the following steps:

1. Bracketing
2. Collecting data
3. Analyzing data
4. Transforming the data
5. Sharing the story (Lemon & Taylor, 1997; Giorgi, 1985; Colaizzi, 1978)

Bracketing

Bracketing means that researchers first attempt to set aside their own beliefs, thoughts, and preconceived notions about the phenomenon under investigation (Byrne, 2001). Researchers bracket so that they can learn about the participants' perceptions without including their own ideas in the study. Lemon and Taylor (1997) describe this process as suspending your previous knowledge of a phenomenon by engaging in a deep self-reflection. As Lancy (1993) suggests, researchers should avoid making assumptions in this type of study. They must enter the process with an open mind to learn the true complexities of a phenomenon.

We suggest writing down your thoughts about the phenomenon. The writing process is closely linked to the thought process, so it is a great way to bracket your perceptions. At the very least, if you identify your biases ahead of time you will recognize them if they enter the process of data analysis. Take time to search your beliefs and write why the topic is important to you, what you think you know about it, and what experiences you have with it.

Data Collection

Interviewing is the most critical and noteworthy method of data collection in phenomenology (Wimpenny & Gass, 2000). Researchers collect data from individuals who have experienced the phenomenon, so they must use very purposeful interview criteria. Once they have identified participants, they conduct an interview or series of interviews.

When interviewing participants, you must cover two broad aspects that are based on the works of Moustakas (1994) and summarized by Creswell (2007):

1. How has the participant experienced the phenomenon?

2. What has influenced their perception of and experience with the phenomenon?

Seidman (1991) advocates conducting three 90-minute interviews with participants. Each interview focuses on a different aspect of the experience:

1. Focused life history

2. Details of the experience

3. Reflection on the meaning

I am using Seidman's approach to interviewing, but I am not able to conduct a third interview with one of the participants. How should I proceed?

Use any data you have already collected that contributes to an understanding of the past and present aspects of the phenomenon. Be sure to document your change in procedures.

Analyze and Transform Data

The analyzing and transforming steps of the phenomenological process are different, but difficult to separate. The analysis step typically follows a process similar to the one described in chapter 4. In phenomenology, researchers identify any statements made by the participants that provide information about the phenomenon. Phenomenologists next thematize these statements, which are often called *meaning units*. The goal of organizing the meaning units and creating themes is to manage information that will ultimately enable the researcher to describe the phenomenon.

The transforming step attempts to distill the phenomenon down to its essence, or the meaning of the experience. Researchers first describe the participants' experiences based on the emergent themes. Next, they explain how the participants experienced the phenomenon and what it means to them. To transform the data, researchers capture the expressions and language that participants use to describe their experiences. In this process, researchers must ask, "How can I describe what I have learned in a manner that captures its essence and does justice to the participant's experience, yet presents the general meaning of the phenomenon?" (Todres, 2005).

Share the Story

When entering this final step of the process, researchers must ask and answer this key question: "From the perspective of the participants, how can this phenomenon be described to others?" Researchers must bring the experience to life for the readers with supporting quotes to illustrate the essence of the phenomenon. At this point, researchers must take the results and descriptions back to the participants for review and approval. This practice is very similar to the process of member checking (Colaizzi, 1978).

Products

This form of qualitative research generates a rich and exhaustive description of a given experience and uncovers unknown aspects of the phenomenon. For example, Goodwin and Compton (2004) unveiled a series of paradoxes that occurred in the lives of women with disabilities, including "the belief that a high quality of life is dependent upon physical wellness and independence and what these young women anticipate for their futures as they age" (p. 134).

If your interests lead you toward a phenomenological investigation, we suggest you take the advice of Lancy (1993) and select a complex phenomenon about which little is known and critical questions still exist. An example of a complex topic is the experience of pain, which Dudgeon, Gerrard, Jensen, Rhodes, and Tyler (2002) investigated in their study titled "Physical Disability and the Experience of Chronic Pain." Using a phenomenological method for less complex topics may "capture too much; it is wasteful" (Lancy, 1993, p. 9). In other words, both the researcher and the research would be better served with a different theoretical approach. Therefore, articulate the phenomenon of interest to your colleagues or advisors, who can help you ascertain whether the topic is complex enough to warrant such an intense investigation.

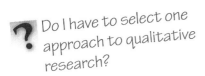 Do I have to select one approach to qualitative research?

No. Although some researchers have stated that using a specific form of inquiry leads to a higher level of sophistication and may "... convey a level of methodological expertise" (Creswell, 2007, p. 232), we respectfully disagree. You can have a substantial level of rigor and obtain a great deal of insight and understanding with a basic or generic interpretive study. Your expertise and the study's sophistication will be apparent in how you plan and execute the study.

Purpose and Questions

Here is a brief example of research questions and purpose statements that would lead a researcher toward a phenomenological study: "The purpose of this...investigation is to describe the essence and structure of learning to make clinical decisions as an athletic training study" (Pitney & Parker, 2002, p. S171). A subsequent research question might be: "How do athletic training students experience and learn clinical decision making?" (Pitney & Parker, 2002, p. S171). Appendix C contains an example of a phenomenological study.

Additional Types of Qualitative Research

This last section focuses on three additional forms of qualitative research. Although they are less common in literature on physical activity and the health professions, you may encounter a qualitative case study, action research study, or narrative inquiry. You should have a basic understanding of each type.

Qualitative Case Studies

Qualitative case studies are intense investigations of a single unit of study, or a bounded system. Examples include one person, one educational program, one school, a single class, a nursing ward, a hospital, or a wellness center.

Merriam (1998) states that researchers may select a bounded system because it is the focus of an identified issue or concern. The bounded system may also be unique in some way. Qualitative case studies use many forms of data to understand what is happening, including observations, documents, and interviews. The product is a rich description of the case and an explanation of how it operates.

Please note that we use the term *qualitative case study* very purposefully. Some case studies, especially in medicine, rely on quantitative data like diagnostic tests to paint a clear picture of what a patient has experienced and how the treatment was effective. Case studies use many forms of data, not just qualitative information. Therefore, you must be very clear about the design you have chosen.

A positive aspect of conducting a case study is that you will not need to work with any participants beyond those involved with the bounded system. In other words, you only need to collect data as it relates to your case. If you decide to conduct a qualitative case study of a unique adapted physical education program for students with multiple sclerosis, you would interview the students, parents, instructors, and administrators. Your observations would be limited to that single program. You would not need to observe other programs or conduct interviews with anyone who is not involved with the case.

A limitation of qualitative case studies is that their findings are rarely transferable. However, you shouldn't feel too apologetic about this limitation. Qualitative case studies tend to focus on unique programs, special concerns, and interesting situations. How they compare to other contexts may be irrelevant; your goal is to vividly describe this case so others can learn about it. Readers can make their own decisions about how to use the findings.

Action Research

Action research is a very practical form of inquiry that is designed to produce a specific outcome, change, or improvement in the very setting in which it is conducted (Stringer, 2004). It is systematically conducted by practitioners, such as clinicians or teachers, to gather information about how procedures are performed (e.g., how teachers teach or how patients are treated) and how context influences outcomes (Mills, 2007; Stringer, 2004). Stringer (2004) elucidated the following characteristics of action research:

- *Change.* Practitioners work on changing their behaviors to improve their practice.
- *Reflection.* Practitioners reflect on, think about, and theorize about their roles and functions.
- *Participation.* Participants change their own practices and behaviors, not those of others.
- *Sharing.* Participants share their perspectives on the context with others.
- *Understanding.* The study enhances understanding of different perspectives.
- *Practice.* Participants apply their findings to their practice.

Action research most commonly employs qualitative data collection and analysis because of the depth of understanding that can be obtained about a specific context. Remember that the focus of action research is changing and improving practice.

Narrative Inquiry

Researchers use narrative inquiry to gain a deep and rich understanding of participants' experiences. The main method is listening to participants' life stories. Narrative researchers believe hearing the stories of others helps you best understand the meaning that participants assign to their experiences.

Because narrative inquiry focuses on life experiences, some researchers tend to link it with life histories, biographies, and even autobiographies (Schram, 2006). Shank (2006) states that narrative inquiry uses strategies of data analysis that are similar to the grounded-theory method, and sometimes employ formal linguistic analysis. However, as Schram (2006) suggests, narrative inquiry is often viewed as running counter to other methods of analysis that break qualitative data down into codes, and then regroup it to form interpretations. When conducting narrative analysis, remember to share your participants' stories in a way that represents what has occurred in their lives. Consider the holistic nature of their stories by presenting lengthy block quotes that capture the meaning of their experiences.

Summary

Many forms of qualitative research exist. The majority of this text addresses the general approach to qualitative research. Researchers who choose a specific type of qualitative research must attend to its foci, methods, and products. Grounded theory examines social processes and seeks to generate a theory or theories. Ethnography focuses on understanding a particular culture. Phenomenology focuses on how participants have experienced a specific phenomenon. Although these three forms of qualitative research are very common, many additional forms exist. Some researchers mix these forms to achieve a unique research purpose.

CONTINUING YOUR EDUCATIONAL JOURNEY

 Learning Through Activity

1. Create a table that lists the forms of qualitative research in a vertical column on the left and the research focus, method, and outcomes in a horizontal row across the top. Fill in the answers for the intersecting cells to create a useful summary.

2. Examine the articles in appendices A, B, and C. Locate the research purpose for each study and identify the form of qualitative research conducted. How could the purpose of each study be changed to warrant using another form of inquiry?

 Checking Your Knowledge

1. This form of qualitative research focuses on understanding an existing culture:
    a. ethnography
    b. grounded theory
    c. phenomenology
    d. a and c
    e. a and b

2. This form of qualitative research produces an exhaustive description of participants' lived experiences:
    a. ethnography
    b. grounded theory
    c. phenomenology
    d. a and c
    e. a and b

3. This form of qualitative research relies on long-term field observation to gain insight and understanding:
    a. ethnography
    b. grounded theory
    c. phenomenology
    d. a and c
    e. a and b

4. This form of qualitative research uses rigorous and systematic procedures of coding to create a set of explanatory concepts:
    a. ethnography
    b. grounded theory
    c. phenomenology
    d. a and c
    e. a and b

5. This form of qualitative research often produces a conceptual model:
    a. ethnography
    b. grounded theory
    c. phenomenology
    d. a and c
    e. a and b

6. Categories are developed in which step of grounded theory?
    a. open coding
    b. axial coding
    c. selective coding
    d. a and c
    e. a and b

7. Which of the following forms of research emphasizes a bounded system?
    a. narrative inquiry
    b. phenomenology
    c. action research
    d. qualitative case study
    e. none of the above

8. A health teacher wants to change a unit on drug and alcohol abuse to raise awareness about the dangers these substances pose for society. What form of inquiry would best serve this purpose?
    a. narrative inquiry
    b. phenomenology
    c. action research
    d. qualitative case study
    e. none of the above

Thinking About It

1. You are planning to study the conflict between the work and home lives of athletic trainers. What form of qualitative research might best achieve this purpose? Why?

2. You are conducting an ethnography to understand how highly competitive sporting environments influence how medical decisions are made. How would you gain access to this environment? How would you collect and analyze data? Whom would you wish to observe and interview? What other forms of data would you wish to collect and analyze?

Making a Stretch

These readings will expand your knowledge on the various forms of qualitative research.

Colaizzi, P.F. (1978). Psychological research as the phenomenologist views it. In R.S. Valle & M. King (Eds.). *Existential-phenomenological alternatives for psychology.* New York: Oxford University Press.

Cortozzi, M. (1993). *Narrative analysis.* Bristol, PA: Falmer Press.

Merriam, S.B. (Ed.). (2002). *Qualitative research in practice: Examples for discussion and analysis.* San Francisco: Jossey-Bass.

Stake, R. (1995). *The art of case study research.* Thousand Oaks, CA: Sage.

Access Colorado State University's online writing guide (http://writing.colostate.edu/guides/research/observe/index.cfm) for more information related to ethnography and narrative inquiry.

Although the appendices provide examples of grounded theory, phenomenology, and ethnographic research, we recommend reviewing these additional examples from other disciplines.

Grounded Theory

Jette, D., Bertoni, A., Coots, R., Johnson, H., McLaughlin, C., & Weisbach, C. (2007). Clinical instructors' perceptions of behaviors that comprise entry-level clinical performance in physical therapist students: A qualitative study. *Physical Therapy, 87*(7), 833-843.

Phenomenology

Schmid, T. (2004). Meanings of creativity within occupational therapy practice. *Australian Occupational Therapy Journal, 51,* 80-88.

Ethnography

Hunter, C.L., Spence, K., McKenna, K., & Iedema, R. (2008). Learning how we learn: An ethnographic study in a neonatal intensive care unit. *Journal of Advanced Nursing, 62*(6), 657–664.

Case Study

McGarvey, H.E., Chambers, M.G., & Boore, J.R.P. (2004). The influence of context on role behaviors of perioperative nurses. *Associate of Operating Room Nurses Journal, 80*(6), 1103-1119.

Action Research

Khunti, K., Stone, M.A., Bankart, J., Sinfield, P., Pancholi, A., Walker, S., et al. (2007). Primary prevention of type-2 diabetes and heart disease: Action research in secondary schools serving an ethnically diverse UK population. *Journal of Public Health, 30*(1), 30-37.

Narrative Analysis

Evans, J., & Penney, D. (2008). Levels on the playing field: The social construction of physical "ability" in the physical education curriculum. *Physical Education & Sport Pedagogy, 13*(1), 31-47.

Being a Good Consumer of Qualitative Research

Learning Objectives

Readers will be able to do the following:

1. Identify three reasons to evaluate qualitative research.
2. Select an appropriate evaluation form or set of questions for each reason.
3. Conduct a thorough and fair evaluation of a qualitative research study.

Evaluating Research

The qualitative community currently has multiple standards for evaluating research that differs according to academic discipline, perspective, and methodology (Creswell, 2007). It is beyond the scope of this chapter to examine each of these in detail, but at the end we provide a list of resources for you to delve into this area of discussion and debate (called criteriology) should you so choose. For now, suffice it to say that qualitative research is evaluated using a variety of standards which differ according to academic discipline, methodology, and perspective. This chapter will help you use appropriate criteria in thoroughly and fairly evaluating qualitative research.

Significance of the Evaluation Process

There are several reasons for evaluating a qualitative study. First, you may need to provide an annotated bibliography that describes each study and analyzes its content. Next, while gathering sources for a review of literature, you may need to determine whether to include a particular study. Finally, you may be asked to review a manuscript for publication or a published study for a research class. In both cases, you will need to conduct a thorough evaluation of the submission. Reviews and evaluations of this nature are critical as they will help you to determine the quality of qualitative research and allow you to become a critical consumer.

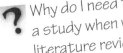 Why do I need to evaluate a study when writing a literature review? Can't I just include the information?

Writing a review is different from writing a summary. Be a critical consumer as you write a literature review, and demonstrate your understanding of the topic by discussing the content and quality of the information presented in the article.

In all three examples, you must determine whether the study in question has been conducted appropriately and rigorously enough to support the findings. For example, you may come across a study with findings that strongly support the importance of an investigation you are proposing. But what if the study you are reading has flaws in design or analysis that cause you to question the results? Or, what if the authors have left out critical information? How will issues such as these affect your use of the study?

Like an investigative reporter, your role is to dissect the study to see whether the critical components are all present and cohesive. You learned about the foundation of this process in chapter 2, which outlines the components of a qualitative research study and shows you how to record them succinctly and accurately. The critical difference between recording and evaluating is one of judgment. Up until this point, you have recorded the contents of a study without passing judgment on their merit. When you record information from an article, you outline what you know about the study. Evaluation often begins with the question, "What do you wish you knew about this study?" In other words, what are the missing pieces? In this context, the missing pieces may be either concrete, such as detailed demographics of the participants, or conceptual, such as how authors reached their conclusions based on the results. Remember, good research often generates more questions than it answers, but the questions should be related to a future study rather than the current article.

Importance of Developing an Impartial Evaluative Lens

As you begin your evaluation, remember that the authors invested time and energy in the process of qualitative research and did not deliberately set out to confuse you! This comment may seem a little flippant, but our intent is to encourage you to approach evaluation with respect for the authors and the process. Think of evaluation as an educational opportunity rather than a negative experience. Evaluating the article will not only assist you in your assignments, it will also help you to strengthen your own writing. Just as reviewing the literature helps you to develop study proposals, evaluating articles helps you determine what works for you as both a reader and a writer.

 I don't mean to be pejorative, but the study I am evaluating seems pathetic. What should I do?

Pathetic is a very strong word. Use it cautiously! You may need to change your evaluative lens for this particular study. Think about how the process of learning from the authors' mistakes strengthens your own research skills.

If you are required to give feedback to this author, you can always find something positive; examples could be their willingness to submit their work for critical evaluation (which takes courage) or the nature of the topic itself for example.

Remember that the content of an article is often at the mercy of journal format, page restrictions, and journal reviewers. If you need additional information, rather than mentally scolding the authors for their lack of detail, consider personally contacting them to ask for clarification. This bold move serves two key purposes. You will both obtain the information you need for your own research and establish contact with a researcher who shares your interests or methodology. Congratulations, you have just begun to develop your research network!

Finally, try to remove any bias about a particular topic or researcher from your evaluation. Just as authors must disclose their biases and research connections, so must you—your evaluation lens must be clear. For example, if you have strongly held beliefs about stem cell research, can you fairly evaluate a study that investigates this topic? In addition, if you have read several articles by the same author and do not particularly like her style of writing or have questioned some of her past research, can you be impartial in evaluating her latest research contribution? Keep these questions in mind as you begin the evaluation process.

Starting the Process

Because several reasons for evaluating qualitative studies exist, your purpose will determine the questions you ask and the time you need to devote to the process. At the most basic level, the evaluation process requires you to judge the "trustworthiness and plausibility of the researcher's account" (Horsburgh, 2003, p. 308). In other words, do you believe that the researchers did what they outlined in their methods section? Do you believe the findings of the study? If so, why do you believe the researchers? If not, why do you distrust them? By answering these questions, you are well on your way to becoming an evaluator!

 My colleague and I read the same qualitative research study, and he thought it was very believable, yet I don't buy the results—that is to say, the study does not seem believable to me. Am I wrong?

Not necessarily. Remember that qualitative research is highly contextual so your interpretation is filtered through your own experiences. The beauty and the frustration of qualitative research is that a study may resonate with your colleague but not with you.

While these question are certainly important, Mays and Pope (1995) identify two goals for qualitative researchers: "to create an account of method and data which can stand independently so that another trained researcher could [analyze] the same data in the same way and come to essentially the same conclusions; and to produce a plausible and coherent explanation of the phenomenon under scrutiny" (p. 109). These goals can be translated into questions for evaluating a qualitative study. Namely, do you have enough information about the study's design and participants to conduct a similar study? Do the authors logically link the results to the conclusions?

As previously mentioned, qualitative researchers are themselves an integral part of the process. Therefore, when evaluating qualitative research you must determine whether the researchers have addressed their personal connection to the context. This phenomenon is called *reflexivity* (Horsburgh, 2003). As you evaluate an article, ask yourself how the author addresses issues of reflexivity.

Finally, evaluations are incomplete without a consideration of the participants. After reading the article, you should feel that you know the participants and understand their perspectives on the phenomenon or experience being studied. To address this issue we suggest that you refer to the explanation of voice emphasis in chapter 8. Which quadrant of voice presentation do the authors use? Is the balance of voices appropriate for this particular study?

As you initially evaluate a qualitative research study, we suggest that you ask the following questions, which we call the *fab five*:

1. Do you believe that the researchers did what they outlined in their methods section? Do you believe the findings of the study? Explain your answer.

2. Do you have enough information about the study's design and participants to conduct a similar study?

3. Do the authors make logical links between the results and conclusions sections?

4. How do the authors address issues of reflexivity?

5. Which quadrant of voice emphasis do the authors use? Does the method help you get to know the participants?

 I don't believe this study, but my advisor is telling me to include it in my review of literature which would change the rationale for my own study. What should I do?

Search your beliefs for the answer. Ask yourself honestly whether you are discounting the study because of its flaws or because it does not support your own research agenda. If there is any truth in the latter answer, set your own agenda aside and do the right thing by including the study.

Digging Deeper

If you are conducting a more thorough evaluation of a study to determine how it should be considered in your review of literature, for example (in addition to the questions listed above), there are others you will need to consider. The key here is to dig a little deeper into the study. Refer back to the sections of chapters 7 and 8 that outline how to write various sections of a qualitative research study. To assist you, we have formulated some questions for evaluation that we call our *top ten*:

1. Does the background information in the introduction lead you logically to the problem statement? Is the research problem related to the purpose statement?

2. Do the authors clearly articulate the study's conceptual framework? Does the literature review support the need for the current study?

3. Have the authors clarified their stance, bias, and experience with the topic under investigation?

4. Do the authors clearly describe the participants and how they were selected?

5. Do the authors fully describe their procedures for data collection and analysis? Are the procedures appropriate for this study?

6. Do the researchers directly and appropriately address issues of trustworthiness?

7. Do the authors present the results clearly and support them with quotes and descriptions?

8. Do the authors compare the key findings of the study with those from existing literature?

9. Are the conclusions logically based on the data presented?

10. Do the authors adequately address the study's limitations?

As you can see from the list, these questions build on our *fab five* and provide more guidance for conducting your evaluation. However, they may make you wonder how many times you can answer no to the preceding questions before you decide to discard a particular study. Fowkes and Fulton (1991) answer this question well when they say, "Unfortunately, there is no magical formula which will convert assessments of details into an overall score on the worth of a paper" (p. 1138). When in doubt, return to the first question of the *fab five*. If you truly do not believe the results and have substantive evidence that supports your perspective, you may need to exclude the article from your literature review. However, attach your notes to the article, keep it in your files, and be prepared to explain your decision.

Conducting a Full Review

Conducting a full review of a manuscript is not a task to be taken lightly. A detailed review can take many hours. Begin by reading the manuscript all the way through, just as the authors intended. This practice serves two key purposes. First, it gives you an overall sense of the paper. The authors deliberately structured their article or manuscript in a particular manner, and you should respect their choice. Second, this process saves you time in the long run. When we were first-time reviewers, we sat down with a list of evaluation questions and filled them out as we read the paper. Although this approach may sound logical, it was often frustrating. We would spend lots of time giving extensive feedback on a piece of information we thought was missing, only to find that very piece of information in a place we had not anticipated. However, it was located exactly where the authors intended. If we had read the entire manuscript, we would have known!

After reading the entire article or manuscript, you must ask some pointed questions. Several authors provide detailed lists for readers to consider when critically evaluating a research report. They include questions related to content (the research itself) and questions related to writing (how the research is presented). We have drawn from the work of Creswell (2007); Locke, Silverman, and Spirduso (1998); McMillan and Wergin (2005); and Merriam (2002) to create a list of questions that we call our *thought-provoking 34!* We hope they will guide you through even the most detailed evaluations.

Title and Abstract

1. Does the title indicate important constructs and relationships from the study?
2. Does the abstract provide enough information to help readers decide whether to examine the full report?
3. After reading the title, abstract, and statement of the purpose, are you interested in reading the next part of the study?

Introduction

4. Do the authors introduce the topic of the study in terms of previous investigations?
5. Do the authors sufficiently explain how the study fits into the present body of knowledge?
6. Is the purpose of the study clearly stated? Does the introductory material make it easy to locate?
7. If the relevant literature contains conflicting findings, did they authors discuss them?
8. Do the references include articles that have been peer-reviewed and published in the last five years?
9. Do the authors clearly define unique terms?
10. Do the researchers provide a balanced view of the problem and make their assumptions clear in the introduction?

Participants

11. Do the authors fully describe the participants?
12. Do the authors fully explain the sampling strategies?
13. Do the authors describe where they observed or interviewed the participants?
14. Do the authors describe why they chose the participants? Do they explain any limitations or potential biases that affected the selection process?
15. Is there evidence that the authors treated participants according to ethical standards?
16. Have the authors explained why the number of participants is appropriate for the study's procedures and purposes?

Methods

17. Do the authors identify and explain the study's design?
18. Do they rationalize their design selection?
19. Do they explain their decisions about the study's design and procedures in terms of the effectiveness or ineffectiveness of previous investigations?

20. Do the authors name and describe all the processes of data collection and analysis used in the study?

21. Do they clearly establish trustworthiness?

22. Do they articulate the environmental conditions in which they collected data?

23. Do the authors indicate which protocol they used for each data collection and analysis procedure?

24. Do they clearly identify their time line for the study and data collection and analysis procedures?

25. Do they articulate how they recorded their data?

Results and Discussion Sections

26. Do the authors clearly report the study's findings?

27. Do they clearly link the findings to the original research questions?

28. Do the titles of the themes capture the essence of the data? Are the themes supported with direct participant quotes?

29. If present, do the figures and tables enhance your understanding of the study?

30. Do the titles and captions accurately represent the content of the figures and tables?

31. Are the authors' conclusions in line with the findings or do they veer into unsupported speculation?

32. Do the authors explain how the study's results fit into the existing base of knowledge?

33. Do the authors thoroughly present the study's limitations?

34. Have the authors adequately answered the question, "So what?"

Personally, when we are asked to review manuscripts for publication, we use this checklist to formulate feedback for the authors. Many journals provide reviewing guidelines with sections for evaluators to complete, but these sections are often broad and correspond to the components of an article (introduction, method, results, and discussion). However, the questions from the *thought-provoking 34* serve two roles. They help us make decisions about the manuscript and they lead us to evidence that supports our decision. From there, we can generate very specific feedback for the authors.

At this point, we focus on aggressive praise and gentle suggestion. It is important to let authors know which aspects need more work, but it is even more important to acknowledge when a job is done well. There is always something positive to say, so don't forget to say it! Imagine that you are sitting with the authors as they read your feedback. If you have written a fair, thorough, and professional review, this image should not make you cringe! If you are looking for the door, perhaps you should rephrase some of your evaluation.

? There are so many evaluation questions and so little time! Is there a short cut for conducting a manuscript evaluation for a journal?

No! If you don't have enough time to thoroughly evaluate a manuscript, you should politely decline the invitation to review it. Remember the quality of your review directly influences the quality of the publication and your reputation as a reviewer.

Qualitative studies must be subjected to thorough and fair evaluation to ensure rigor and integrity of research. The following figure represents the information on evaluation presented thus far:

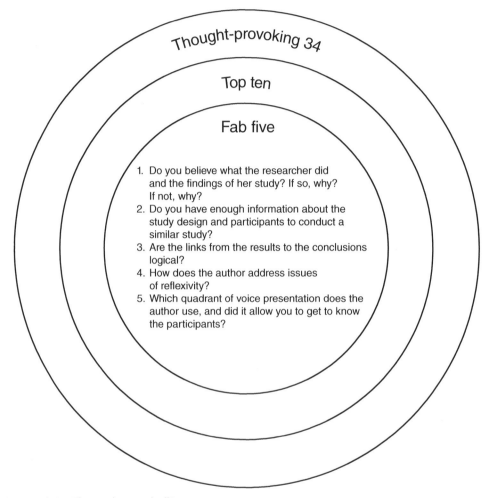

Figure 10.1 The evaluation bull's-eye.

Think of the evaluation process as a target in which the questions from the *fab five* form the bull's-eye, and the questions from the *top ten* and *thought-provoking 34* form the concentric outer circles. Your starting point depends on your reason for evaluation, but you will always need to return to the center.

Recognizing Specific Types of Qualitative Research

This chapter provides three sets of questions to guide you through the evaluation process of general, interpretive qualitative research. In chapter 9, you learned about three different types of qualitative research: grounded theory, ethnography, and phenomenology. If the research report you are evaluating uses one of these specific approaches, you may need to ask additional questions. Before you read this following paragraph, remind yourself of the critical elements of each approach. These elements will drive the evaluation questions.

Grounded Theory

When evaluating a grounded-theory study, look for a clearly defined link between the themes of the data and the theoretical model presented. If either component or the link itself is missing, you should question the study. Creswell (2007) suggests that a figure or diagrammatic representation of the theory should also be present in the article. Combined, these two ideas may be expressed in the following evaluation questions:

- Do the authors clearly articulate the connection between emergent data categories and the theoretical model?
- Is the theoretical model explained with both text and graphics?

Ethnography

An ethnography focuses on developing an understanding of an entire culture or cultural group. Therefore, ethnographers immerse themselves in other cultures for extended periods of time. When evaluating an ethnographic study, ask yourself if the researcher has spent enough time immersed in the culture to provide a rich description and interpretation?

Phenomenology

The challenge of a phenomenological study is to describe the essence of a common experience. The researcher must describe the experience and what it means. When evaluating a phenomenological study, ask yourself whether the authors convey the overall experiences of the participants and describe the experience and the context (Creswell, 2007).

Summary

Before it can claim a place in the research community, a qualitative study must be fairly evaluated to determine its rigor and plausibility. This chapter outlines three layers of the evaluation process. Your starting point depends on your reason for evaluating a qualitative study. Finally, this chapter provides some questions for you to consider if the study you are evaluating is either a grounded theory, ethnography, or a phenomenology.

CONTINUING YOUR EDUCATIONAL JOURNEY

 Learning Through Activity

1. In chapter 2, you read and recorded information from a study by Pitney (2002). Take the next step by completing a critical evaluation of the same study using the questions in the *thought-provoking 34*.

2. Use the information from your evaluation to write three paragraphs of feedback for Pitney.

 Checking Your Knowledge

1. The debate about whether consistent and common criteria should exist for evaluating qualitative research is called
   a. criteriology
   b. terminology
   c. phenomenology
   d. radiology
   e. evaluometrics

2. At the most basic level, when evaluating a manuscript, you must pass judgment on which of the following aspects?
   a. trustworthiness
   b. plausibility of the researcher's account
   c. length of the manuscript
   d. a and b

3. This term denotes whether researchers have disclosed their biases and personal connections with a topic of study:
   a. reflexology
   b. reflexivity
   c. self-disclosure
   d. a and c
   e. a and b

4. When conducting a full manuscript review, your first step should be to
   a. provide feedback on the introduction
   b. read the entire manuscript
   c. give specific feedback on the methods
   d. make an initial judgment of the topic's importance to you
   e. a and b

5. When reviewing this form of research, you must determine whether the researcher has spent enough time immersed in the culture:
   a. ethnography
   b. grounded theory
   c. phenomenology
   d. a and c
   e. a and b

6. Which of the following should be clearly identified in the manuscript of a grounded-theory study?
   a. cultural description

b. graphic image with a time line for the study

c. exhaustive description of the phenomenon

d. theoretic model

e. a and b

Thinking About It

1. You have been asked by a journal to conduct a blind review of a manuscript, but you can identify the author by its title. How would you handle this situation?

2. A colleague is very upset by a review she received for her manuscript and asks for your opinion. When you read the review, you realize that you wrote it! How should you proceed?

3. You are angry because you do not agree with the committee's evaluation of your thesis draft. You plan to either write a scathing response or call a committee meeting. Is this your best choice of action? What other options might you have?

Making a Stretch

The following resources provide an interesting dialogue on some of the issues associated with evaluating qualitative research.

Creswell, J.W. (2007). *Qualitative inquiry and research design: Choosing among five approaches* (2nd ed.). Thousand Oaks, CA: Sage.

Lazarton, A. (2003). Evaluative criteria for qualitative research in applied linguistics: Whose criteria and whose research? *The Modern Language Journal, 87,* 2–12.

Lincoln, Y.S. (1995). Emerging criteria for quality in qualitative and interpretive research. *Qualitative Inquiry, 1,* 275–289.

Richardson, L., & St. Pierre, E.A. (2005). Writing: A method of inquiry. In N.K. Denzin & Y.S. Lincoln (Eds.), *The Sage handbook of qualitative research* (3rd ed., pp. 959–978). Thousand Oaks, CA: Sage.

Sparkes, A.C. (2001). Myth 94: Qualitative health researchers will agree about validity. *Qualitative Health Research, 11*(4), 538-552.

Looking Back and Moving Forward

Learning Objectives

Readers will be able to do the following:

1. Recognize and respond to arguments against qualitative research.
2. Explain when to use mixed methods for a study.
3. Identify practical resources for conducting qualitative research.

Looking Back

The preceding chapters cover a great deal of information about qualitative research, including how to recognize, conceptualize, and execute a qualitative study. Our intent was to write a comprehensive text that provides practical examples and describes the systematic processes of qualitative research in language that is easy to understand. Metaphorically, we wanted to provide a view of the forest without interference from the trees. Having read the first 10 chapters, you are ready to walk into the forest of qualitative inquiry. Our final task is to provide tools for the journey of effective qualitative research.

You may wonder what you will need to be a successful qualitative researcher. We present three sets of tools to assist you. The first will help you respond professionally to frequent arguments made against qualitative research. The second will help you understand situations that require using other methods in tandem with qualitative practices to address specific research questions. Finally, the third provides practical, nuts-and-bolts information about completing a qualitative study.

Defending Your Qualitative Research

This section begins with three anecdotes that illustrate some of the biases against qualitative research that we have experienced personally. William Pitney was once asked at a job interview if he did "that touchy-feely research." Despite the fact that one of us has an undergraduate minor in mathematics, we have been asked if we are afraid of the statistics associated with quantitative research. We as qualitative researchers have also been called the junior varsity team compared to the varsity team of quantitative researchers.

Although these examples are meant to provide levity, they elucidate a dominant scholarly attitude toward the interpretive research paradigm. You will likely encounter critics during your work in qualitative research, and you could easily feel defensive when articulating your position. However, emotional arguments are usually unproductive and unpersuasive. In most instances arguments and criticisms are borne from myths that are created by assumptions that are held about the interpretive research paradigm. This section outlines typical assumptions against and criticisms of qualitative research and suggests potential responses. The names of the two characters, Dr. Skeptic and Dr. Convincing, are deliberately chosen to signify the positions of the quantitative and qualitative researchers. You have a professional obligation to politely respond to Dr. Skeptic. However, you should not be subjected to Dr. Angry, Dr. Pejorative, or Dr. Disrespectful. When you meet Dr. Skeptic, be strong. Do not become Dr. Defensive or Dr. Emotional!

? *Why are some scholars skeptical of qualitative research in light of the strong arguments supporting its value and use?*

Good question! You will encounter colleagues who are only familiar with the quantitative paradigm. With such a focused approach to research, being open to and accepting alternative methods can be challenging. Our suggestion is to keep gently educating where possible, and remember that change takes time, but it is always worth the wait.

Assumption 1: Qualitative Research Is Not Rigorous

Dr. Skeptic: I am really surprised that you have chosen to conduct qualitative research. It appears that you are taking the easy way out. Qualitative research is simply not rigorous.

Dr. Convincing: If you determine rigor primarily with the standards of reliability and validity, I can understand your concern. Honestly, I also feel skeptical anytime I read a study that does not address why I should trust the data presented. However, a thorough process similar to that of quantitative research exists within the interpretive paradigm called *establishing trustworthiness*. Trustworthiness is a general term that covers issues of credibility, transferability, and dependability of data. I would be happy to share specific strategies that qualitative researchers use to ensure trustworthiness with you.

Assumption 2: Qualitative Research Is Not Objective

Dr. Skeptic: Qualitative research seems so subjective and biased. Most of the data includes personal perceptions of experience. Objective data are far more important, are they not?

Dr. Convincing: Some scholars prefer data that can be quantified or measured. Many disciplines require measurements to answer specific questions, such as how inflamed tissue responds to medication. However, not all research questions can be answered with measurements. This is particularly true in social sciences, which attempt to understand human relationships and experiences. Subjective data, such as beliefs, perceptions, values, and attitudes, influence how humans act and interact.

The concept of subjectivity also relates to the way in which research is conducted. Aren't all research methods of data collection and analysis beholden to the decisions of the researcher? Any time a researcher makes a choice, a level of subjectivity enters the scene. All research methods have strengths and weaknesses, and you must select the approach that best suits your research question. The use of subjective information should not automatically discredit a study.

Assumption 3: Qualitative Research Is Not Scientific

Dr. Skeptic: I believe that the most important and highest form of knowledge is based in science. To me, qualitative research just seems too arbitrary and unsystematic to be considered scientific.

Dr. Convincing: To answer you, I need to refer to the work of Dingwall, who succinctly states, "one of the greatest methodological fallacies of the last half century in social research is the belief that science is a particular set of techniques; it is, rather, a state of mind, or attitude, and the organizational conditions which allow that attitude to be expressed" (1992, p. 163). The hallmark of science is empirical, systematic inquiry to answer purposeful and important research questions. My goal as a qualitative researcher, therefore, is gain insight and understanding through clearly articulated, systematic methods that investigate the perceptions and experiences of others.

Assumption 4: Qualitative Research Is Merely Storytelling

Dr. Skeptic: I really enjoyed the story you wrote about the experiences of high school students with obesity in physical education classes.

Dr. Convincing: What story are you referring to?

Dr. Skeptic: The one you published in the latest issue of our journal.

Dr. Convincing: Oh, that was actually a qualitative research study.

Dr. Skeptic: That was research?

Dr. Convincing: Yes. A story contains several characters and a literary plot. A research study systematically collects, analyzes, and interprets data, then identifies major themes or patterns of participant experience. The study you read presented several themes identifying students' perceptions of their weight and their experiences in physical education programs.

Let me clarify with the example of a standard fairy tale. If the story of the three little pigs were a qualitative research study, it would fully outline the participants' demographics, including the average age of the pigs and the duration of time they spent in their homes. It would also compare the location of the homes to one another and articulate whether the pigs had a previous relationship with the wolf. Data from detailed interviews would then identify the wolf's motive for destroying the homes of the first two pigs. You, as a researcher, might ask the question, "What process did the big bad wolf use to destroy the homes?" You would likely conduct a grounded-theory study to answer that question and create a preventive theoretical model. As you can see, fundamental differences exist between a story and a qualitative research study.

Assumption 5: Qualitative Research Cannot Be Generalized

Dr. Skeptic: What good is qualitative research if I can't apply the findings to a broad population? Doesn't this difficulty in generalization substantially limit this form of study?

Dr. Convincing: The purpose of qualitative research is to gain insight and understanding, not to apply findings to a broad population. Although the findings may not be generalized, they can often be transferred to similar contexts. For optimal transferability, qualitative researchers must carefully describe the details of the context to help readers locate the findings that apply to their own situations. This technique helps qualitative research make a valuable contribution to the existing base of knowledge.

Assumption 6: Qualitative Research Cannot Be Reproduced

Dr. Skeptic: Because the findings of qualitative research cannot be reproduced, I believe they are unreliable. My research would be severely criticized if another researcher could not replicate my study and its findings.

Dr. Convincing: It is very important for quantitative researchers to be able to reproduce their findings, especially in controlled laboratory settings. However, qualitative researchers study participants in their natural settings, which vary greatly from one study to the next. The people, school, and local community may all be different. Therefore, qualitative researchers should not expect to obtain the exact same results when repeating a study. They should explain the steps of their study with enough clarity and detail that another researcher could conduct a similar investigation.

? My advisor recommends taking courses in both quantitative and qualitative research, but I want to focus on the qualitative paradigm. What should I do?

Taking both classes will make you a well-rounded and well-respected researcher.

The preceding examples outline the role of educator that you will likely assume to help others understand qualitative research. Be prepared to provide information about the nature of qualitative research, such as how data are collected and analyzed, and how trustworthiness is addressed. Because you may need to compare and contrast qualitative and quantitative research, you should also learn about the quantitative paradigm. As you speak with others and educate them about qualitative research, you might point out that each research paradigm has a different focus and addresses different problems. In many instances, the paradigms can be used together to thoroughly research a particular topic. This approach is called *mixed methods.*

Combining Methods

Suppose you create a study that has more than one purpose or poses many different research questions. When selecting your approach, you realize that no single method will address the scope of your study. What should you do now? You have two options: you can either narrow the scope of your study for use with a single method or use mixed methods design to address all your research questions.

Mixed methods research has gained popularity in the past decade because of its comprehensive nature. In fact, there is now a scholarly journal dedicated to mixed methods research. This section defines this type of research, explains how it is conducted, and provides some examples of mixed methods studies.

 I have conceptualized my study and developed my purpose and research questions. I think that a mixed methods approach would be the most appropriate, but I have limited knowledge of quantitative research. What should I do?

The answer to this tough question depends on your situation. If you need to conduct the research independently, as in a thesis or dissertation, you may need to narrow your study down to the qualitative component. However, be wary of letting your methods drive your research questions! You have two alternatives. You may either collaborate with a quantitative researcher or learn quantitative methods and conduct your original study alone.

Mixed methods researchers collect and analyze both qualitative and quantitative data to answer specific research questions. They conduct studies in one of the following ways:

1. *QNR to QLR.* They first collect the quantitative data (QNR), and then collect the qualitative data (QLR).
2. *QLR to QNR.* They collect qualitative data before quantitative data.
3. *QLR with QNR.* They collect qualitative and quantitative data concurrently.

The order of data collection depends on the nature of the research and order in which research questions should be answered. Thus, when designing a mixed methods study, researchers must identify which method should address each research question.

Creswell (2003) explains that the nature of the research study, including the researcher's interests, the audience, and the focus of the study, determines the dominant form of inquiry. Consider a study on spirituality in athletic-training curricula completed by Udermann, Schutte, Reineke, Pitney, and Gibson (2008). The authors conducted a survey in which the majority of questions had closed-ended responses (yes or no). However, they concluded the survey with an open-ended question about participants' spiritual views. The dominant method was quantitative, but the less-dominant, qualitative component also explained some of the survey's findings.

Pitney, Stuart, and Parker (2008) conducted a mixed methods study that examined role strain among physical educators who also worked as athletic trainers in the secondary-school setting. They used quantitative methods for research questions related to the prevalence and type of role strain in this population. They used qualitative methods for research questions on the contextual factors that influenced role strain.

The authors first used surveys and a specific instrument that identified the level and type of role strain to collect the quantitative data. They invited survey respondents to participate in the qualitative portion of the study, which involved detailed interviews. The authors chose participants with each level of role strain (low, moderate, and high) for the qualitative portion of the study. In this QNR to QLR example, the qualitative data both supported and extended the survey's findings by fully explaining the factors that contribute to role strain.

Using Practical Considerations

Now that you understand the arguments made against qualitative research and methods of mixing the qualitative and quantitative research paradigms, consider this practical advice. We have observed that when conducting studies, students often overlook small details that could make the process much more successful and enjoyable if included.

Preparing for Data Collection

In order to be efficient and effective, you must carefully prepare to collect data. We suggest that you pack your research bag ahead of time with the following items to be both effective and efficient:

- *Interview or observation guide.* Bring several unmarked copies for recording your notes.

- *Omnidirectional microphone.* This instrument picks up all voices, which is especially critical when conducting focus groups in which sounds come from different directions. Note that this item often has its own battery and on/off switch.

- *Audio-recording device.* Pack plenty of new batteries. Most digital recorders now have a great deal of memory that lets you save more than one interview. If you are using a tape recorder, bring an ample supply of blank audio tapes.

- *Directions to the data-collection site.* This item is self-explanatory! Prepare an alternative route in case of construction or bad traffic.

- *Video-recording device.* This device is helpful for conducting observations. Make sure to obtain permission from participants before recording. Check that the tool has adequate storage capabilities. Pack extension cords and extra batteries.

- *Contact information.* Bring a phone number for your participant or primary contact at the site in case you are late or need to reschedule.

- *Informed consent forms.* Bring several unmarked copies for participants.

- *Clipboard or something sturdy to write on.* Your writing must be legible, and the site may not provide a table or other hard surface.

- *Appropriate attire.* Consider your site and participants, and dress to fit in, not stand out!

- *Blank paper and writing utensil.* You may desire an additional method for recording information.

- *Photo identification.* It is embarrassing and inconvenient to be turned away from your collection site because you lack the proper identification.

I am a graduate student with a very tight budget. Do I need to purchase all my research equipment personally?

Honestly, all research comes at a cost and you must create a budget. However, ask your institution, advisor, and colleagues if they have items you can borrow before you buy anything. Your institution may also have funding available for research expenses. Start asking for help early in the research process!

After packing your bag with the essentials for data collection, consider bringing duplicates of certain items in case something goes wrong. Think of these extra items like insurance: you may not use it when you have it, but when you don't have it, you need it!

- *Batteries.* Make sure they are new and fit your equipment.

- *Recording device.* If possible, bring an extra camera or voice recorder.

- *Pens.* You can never have too many!

- *Consent forms.* You never know whom you might meet at the site. You may wish to do additional interviews or recruit new participants.

Your bag is packed and you are ready to go. Before leaving, please consider these tips that we learned from the school of hard knocks!

- *Practice, practice, practice.* Rehearse conducting interviews and setting up your equipment. Setting up always takes longer than you think, but the process will run more smoothly if you know what you're doing.

- *Learn how to use your equipment.* You must thoroughly understand the intricacies of your equipment. Do you have the widget that opens the battery compartment of the camera? Do you know where the camera's battery compartment is? Do you know where to plug in the microphone? Does your equipment have a separate power source?

- *Assign pseudonyms carefully.* If you intend to let participants select their own pseudonyms, you must have a plan for matching the pseudonyms with the participants' names in case you want to conduct additional interviews.

- *Arrange a suitable interview location ahead of time*. Don't assume that your participant will select the best venue for your needs. Visit the interview site ahead of time if possible, or arrive early enough to make necessary adjustments.

Following Up

- *Thank your participants*. Hand-write and send legible thank-you notes to everyone who assisted your process. You can also provide a typed acknowledgement of participation and time commitment for your participants' files or portfolios.

- *Practice self-talk while leaving the site*. Have a recorded conversation with yourself on the way back from the research site to capture your immediate thoughts about your experience. This recording of your initial reactions helps your reflection process.

- *Secure your data*. You will probably need your research bag for data collection again, so check the equipment and then repack it. However, you must remove your data from the bag immediately, back it up or make copies, and then store it in a secure place. You may never leave your data in a place where others could access it, whether intentionally or unintentionally. This is the contract you entered into with your IRB and you must abide by it.

- *Begin transcriptions promptly*. The process of transcribing interviews or typing up observations always takes longer than you think it will, so start early. The longer you wait to begin, the more imposing the task will seem. Again, back your transcriptions up electronically and store them in a secure place. Do not use a hard drive, which could easily crash, as your sole storage unit.

Using Software for Qualitative Research

Chapter 4 explains the general steps of qualitative analysis. Many researchers approach analysis manually by writing coded concepts on index cards and then organizing them into themes or categories. Some researchers use spreadsheets or word-processing files for this process. Others use programs known as Computer Assisted Qualitative Data Analysis Software (CAQDAS) to organize and sort textual data (Webb, 1999). This section provides information on common software. However, we must emphasize that although software helps you code, organize, store, and retrieve your data, it will not interpret it for you (Morison & Moir, 1998). The researcher is the instrument for data collection and analysis. Webb suggests that novice qualitative researchers use a manual technique rather than a CAQDAS program (1999). Once you have gained experience with the interpretive process, you may progress to software programs.

Many programs exist, but we share the following three programs here because either we or our colleagues have successfully used them. We don't believe that any one program is better than all the others. Because each program has particular nuances and ways of organizing data, you must choose the one you feel most comfortable with. Although choice of software is largely a matter of personal preference, graduate students and new faculty members usually use the programs made available by their departments. Consider reading Barry's online article (1998) comparing two software packages at www.socresonline.org.uk/3/3/4.html. Although this article is more than 10 years old, it provides some helpful considerations for choosing software. We also recommend Alexa and Zuell's (2000) review and analysis of 15 different software packages.

- *The Ethnograph.* The Ethnograph is an early form of analysis software that helps researchers import their interview transcripts from a word-processing program. The software helps researchers organize data with a code book or code families. Go to www.qualisresearch.com for more information.

- *ATLAS.ti.* ATLAS.ti also helps researchers import and organize textual data. The program displays transcripts and allows the researcher to code important pieces of information. It also has a feature for importing and storing audio and video files, such as recordings of observations. Another interesting attribute of the program is the concept-mapping feature, which allows researchers to explore the relationships between the data's themes. Go to www.atlasti.com for more information.

- *NVIVO.* NVivo software allows researchers to import, code, and organize transcripts and documents that contain tables and figures. The program's features let researchers keep memos and display relationships between coded concepts. Go to www.qsrinternational.com//products_nvivo.aspx for more information.

? I don't have time to simultaneously conduct my study and learn to use computerized software. What should I do?

You should never attempt these two tasks at the same time! You should master a computerized software package before you even turn on a microphone or begin collecting data. If this is not the case, then manually analyzing your data will actually be more efficient.

Locating Resources

The following text-based resources will help you answer questions as you move forward in your journey with qualitative research:

- Denzin, N.K., & Lincoln, Y.S. (Eds). (2000). *Handbook of qualitative research* (2nd ed.). Thousand Oaks, CA: Sage.

  If ever you need a tome with information on every qualitative research topic, choose this one. This comprehensive text is written by a number of highly respected authors. Its topics include the historical perspective of qualitative research, ethics and politics, and performing qualitative research.

- Patton, M.Q. (2002). *Qualitative research & evaluation methods* (3rd ed.). Thousand Oaks, CA: Sage.

  This wonderfully comprehensive text is easy to read, practical, accessible, and well-organized.

- Schwandt, T. A. (1997). *Qualitative inquiry: A dictionary of terms.* Thousand Oaks, CA: Sage.

  What better resource to have than a dictionary dedicated to qualitative research? As you continue your research journey, this dictionary helps clarify many terms you encounter in qualitative research.

Consider these computer-based resources as well:

- *The Qualitative Report.* This online journal has many articles and links, as well as its own search engine. The links are organized with information on practicing, teaching, and assessing quality of qualitative research. Go to www.nova.edu/ssss/QR/index.html to access the journal.

- *The Qual Page.* The Qual Page contains many resources related to qualitative research, including methods, theses, reports, and philosophical foundations. Go to www.qualitativeresearch.uga.edu/QualPage/ to access the site.

- *The Cochrane Qualitative Research Methods Group.* The Cochrane Qualitative Research Methods Group (CQRMG) has a Web site to develop awareness of qualitative research for health care professions. The CQRMG site explains the role of qualitative evidence in health care and provides information for reviewing qualitative research. Go to www.joannabriggs.edu.au/cqrmg/index.html to access the site.

Summary

This chapter provides closing advice, including how to counter arguments against qualitative research, how to conduct mixed methods studies, and how to successfully collect data. Embrace the role of educator and advocate for qualitative research. Consider the practical necessities of your task as researcher. Access and read about recent developments in the research community. Finally, prepare to enjoy your future as a qualitative researcher!

CONTINUING YOUR EDUCATIONAL JOURNEY

 Learning Through Activity

1. Conduct an Internet search for useful information that has been updated in the past year using the following key terms: *qualitative research, qualitative data analysis,* and *interpretive research.*

2. Find a mixed methods research study (other than those previously cited) and determine whether the quantitative or qualitative method was dominant. Explain how the results from each method complemented one another.

3. Search the Internet for information about two software packages for qualitative research other than those previously cited. Compare and contrast the programs.

 Checking Your Knowledge

1. Which of the following is a common misconception about qualitative research?

   a. It is not rigorous.

   b. It is scientifically based.

   c. It gains insight and understanding.

   d. It is objective.

   e. b and c

2. In a mixed methods study, the qualitative component must always follow the quantitative component.

   a. true

   b. false

3. In a mixed methods study, the qualitative component is always dominant.

   a. true

   b. false

4. The beauty of a qualitative research software package is that it will analyze the data for you.

   a. true

   b. false

5. After collecting data, your first step must be to secure and back up your files.

   a. true

   b. false

Thinking About It

1. You are debating the value of qualitative research with a colleague. After some time, you realize that no matter how you articulate your stance, your colleague will always consider your research inferior. How do you manage this situation?

2. You have conducted a mixed methods study and submitted your manuscript for review. Your reviewer considers mixed methods studies weaker than pure, single method studies. How should you respond?

3. After arriving at your interview site, you realize you forgot your informed consent forms. What options do you have now?

Making a Stretch

These readings will expand your knowledge on using qualitative research software.

Alexa, M., & Zuell, C. (2000). Text analysis software: Commonalities, differences, and limitations: The results of a review. *Quality & Quantity, 34*, 299-321.

Lewins, A., & Silve, C. (2007). *Using software in qualitative research: A step-by-step guide.* Thousand Oaks, CA: Sage.

Pope, C., Ziebland, S., & Mays, N. (2000). Qualitative research in health care: Analysing qualitative data. *British Medical Journal, 320*, 114-116.

Webb, C. (1999). Analysing qualitative data: Computerized and other approaches. *Journal of Advanced Nursing, 29*, 323-330.

Appendix A

The Professional Socialization of Certified Athletic Trainers in High School Settings: A Grounded Theory Investigation

Northern Illinois University, Dekalb, IL

William A. Pitney, EdD, ATC/L, provided conception and design; analysis and interpretation of the data; and drafting, critical revision, and final approval of the article.

Address correspondence to William A. Pitney, EdD, ATC/L, Department of Kinesiology and Physical Education, Northern Illinois University, Dekalb, IL 60115. Address e-mail to wpitney@niu.edu.

Objective: To gain an understanding of the professional socialization experiences of certified athletic trainers (ATCs) working in the high school setting.

Design and Setting: A qualitative investigation using a grounded theory approach was conducted to explore the experiences related to how ATCs learned their professional role in the high school setting.

Participants: A total of 15 individuals (12 ATCs currently practicing at the high school level, 2 current high school athletic directors who are also ATCs, and 1 former high school ATC) participated in the study. The average number of years in their current position for the 12 currently practicing ATCs was 10.16 ± 7.44, with a range of 2 to 25 years. The 2 athletic directors averaged 5.5 years of experience in their roles, and the former high school athletic trainer had worked in that setting for 1 academic year.

Data Analysis: The interviews were transcribed and then analyzed using open, axial, and selective coding. Peer debriefing, member checks, and triangulation were used to establish the trustworthiness of the study.

Results: Informal learning processes were discovered as the overarching theme. This overarching theme was constructed from 2 thematic categories that emerged from the investigation: (1) an informal induction process: aspects of organizational learning, and (2) creating networks for learning.

Conclusions: Informal learning is critical to the professional socialization process of ATCs working in the high school setting. Because informal learning hinges on self-direction, self-

Entire text of appendix A is reprinted, by permission, from W.A. Pitney, 2002, "The professional socialization of certified athletic trainers in high school setting: A grounded theory investigation," Journal of Athletic Training 37(3): 286-292.

evaluation, reflection, and critical thinking, the findings of this study indicate that both pre-service and continuing education should attempt to foster and enhance these qualities.

Key Words: informal role induction, learning networks, organizational socialization, self-evaluation, reflection, critical thinking

Professional socialization is a process by which individuals learn the knowledge, skills, values, roles, and attitudes associated with their professional responsibilities.[1] Socialization is considered to be a key component of professional preparation and continued development in health and allied medical disciplines[2,3] and has been investigated in medical education,[4,5] nursing,[6,7] occupational therapy,[8] and physical therapy.[9]

Professional socialization is typically exemplified as a 2-part developmental process that includes experiences before entering a work setting (anticipatory socialization) and experiences after entering a work setting (organizational socialization).[10] The first process, anticipatory socialization, refers to experiences such as one's formal training as an undergraduate or graduate student, background as an employee in another setting such as an Emergency Medical Technician, or prior experience as a volunteer with an organization such as the American Red Cross. Organizational socialization refers to experiences such as in-service education and mentoring. The organizational socialization phase of professional socialization can be divided into 2 parts; (1) a period of induction, and (2) role continuance.[10] Induction experiences take many forms. For example, induction processes can be either very formalized (ie, requiring employees to attend specific orientations or instructional sessions) or very informal (ie, no orientation). Additionally, induction processes may be either sequential, requiring specific skills to be learned at specific times during the initial periods of a job, or random, having no time frame for the development of various skills within the organization. Role continuance, on the other hand, focuses on adjusting to the organizational demands over time and continually learning the nuances of a given role and developing professionally. While athletic training has given a great deal of attention to the anticipatory socialization by way of the professional preparation process, there is a paucity of research related to organizational socialization.

Organizational socialization relates to how individuals adapt to their new roles and learn about what is acceptable practice in dealing with the demands of their work. For example, the organizational socialization can be very structured, such as having an athletic director orient a new employee in a very systematic manner, or this process can be unstructured, leaving the employee to ask questions of other employees as various situations arise. Understanding the organizational aspects of professional socialization allows the discovery of the necessary aspects of professional development in a work setting and can serve to improve both athletic training education and continuing education strategies. The purpose of this study, therefore, was to explore the professional socialization of certified athletic trainers (ATCs) in the high school setting in order to gain insight and understanding into how they initially learned and continued to learn their professional responsibilities in an organizational setting.

Methods

Because the purpose of the project required an exploration of the actual experiences of ATCs in the high school setting and how the setting influenced their learning, qualitative methods were employed. The fundamental objective of qualitative research is to gain insight into and understand the meaning of a particular experience,[11,12] and the context that influences the meaning,[13] rather than drawing firm conclusions. Qualitative research is also well suited to study processes[13] such as professional socialization.

In qualitative research, the protection of the participants' anonymity is paramount. Therefore, audiotape recordings of interviews were transcribed and labeled with pseudonyms that are used at various locations in the manuscript. Moreover, at the completion of the study, the audiotapes were destroyed, but the transcripts were maintained using the established pseudonyms. Before data collection, appropriate institutional review board approval was received.

With qualitative methods, the researcher is the primary "instrument" for data collection and analysis, and extreme sensitivity is given to the nature and perspectives of human participants. A researcher's perspective, however, can shape the analysis and interpretation of the qualitative data. My perspectives about the high school setting were shaped in 2 ways. First, I was formerly employed as a clinical high school ATC and interacted with coaches, athletic directors, and athletes and their parents. Second, at the time of data collection and analysis, I was a faculty member in a Commission on Accreditation of Allied Health Education Programs (CAAHEP)-accredited program that used several high school sites as clinical education experiences for the athletic training students. Entering this study, I believed that professionals are not simply products of their work environments but rather active participants in their professional development and that this process continues throughout their careers. This line of thought is consistent with symbolic interactionism, which is often used as a theoretic basis for grounded theory studies.[14]

Participants

A total of 15 individuals (12 ATCs currently practicing at the high school level, 2 current high school athletic directors who are also ATCs, and 1 former high school ATC) participated in the study. The average number of years in their current position for the 12 currently practicing ATCs was 10.1 ± 7.44, with a range of 2 to 25 years. The 2 athletic directors averaged 5.5 years of experience, and the former high school ATC had worked in that particular setting for 1 year before entering graduate school. The athletic directors and former high school ATC were included in order to cross-reference, or triangulate, the perspectives of the currently practicing ATCs. Six participants were women and 9 were men. The participants represented 3 different midwestern states. Participants were initially purposefully selected: that is, I recruited volunteers whom I knew and who agreed to interviews. I then asked these individuals for suggestions of other ATCs who might be willing to participate. The remaining individuals were contacted via either e-mail or phone and agreed to interviews. Before the interviews, participants were required to review and sign an informed consent form. The Table identifies other pertinent participant demographic information.

Data Collection and Analysis

Data were collected using semistructured interviews. Each interview incorporated several key questions or open-ended statements, including the following:

1. Describe your first few years of being an ATC at the high school level.

Participants' Demographic Information and ATC* Experience

PARTICIPANT PSEUDONYM	SEX	SCHOOL ENROLLMENT†	CITY POPULATION‡	CURRENT POSITION	YEARS IN CURRENT POSITION	PREVIOUS EXPERIENCE AS ATC
Betty	F	3000–3500	100 000–115 000	Athletic director	5	High school setting
Marsha	F	2000–2500	60 000–75 000	Athletic director	6	High school setting, clinical setting
Reginald	M	2000–2500	45 000–60 000	Teacher/ATC	25	College setting
Claire	F	1000–1500	<15 000	Administration/Head ATC	8.5	College setting
Donald	M	2000–2500	30 000–45 000	Full-time head ATC	23	Graduate assistant, high school setting
Payton	M	1000–1500	<15 000	College instructor	n/a	None before entering high school setting, first position after bachelor's degree
Jennifer	F	1000–1500	30 000–45 000	Teacher/Head ATC	9	Graduate assistant, high school setting; assistant ATC, teaching aide
Douglas	M	2000–2500	45 000–60 000	Teacher/Head ATC	9	Clinical setting, high school setting
Alicia	F	2500–3000	15 000–30 000	Full-time head ATC	2	None
Robert	M	2500–3000	115 000–130 000	Full-time head ATC	16	College setting, clinical setting
Theo	M	1500–2000	30 000–45 000	Teacher/Head ATC	4	None
Jeremy	M	1500–2000	15 000–30 000	Full-time ATC	8.5	Clinical setting, high school setting
Carol	F	2000–2500	45 000–60 000	Teacher assistant/ATC	9	None
Samuel	M	1500–2000	15 000–30 000	Teacher/ATC	4	Clinical setting
Beck	M	3500–4000	< 15 000	Teacher/ATC	4	Full-time teaching, no ATC responsibilities for 1 year

*ATC indicates certified athletic trainer.

†Data obtained from the appropriate state high school athletic association Web site.

‡Each city was located in a metropolitan area as designated by the US Census Bureau; population data obtained from the US Census Bureau.

2. How did you learn your role and professional responsibilities at the high school level?

3. What has been your greatest challenge at the high school level, and how did you learn to deal with it?

4. What do you like best, or what are the good things about your current position?

5. What aspect of your job do you feel least satisfied by?

6. What is, or how would you describe, your professional mission?

7. What motivates you on a daily basis?

8. What advice might you give to an ATC just about to enter the high school setting for the first time?

Because both athletic directors were former ATCs practicing in the high school setting, they were asked to reflect on their experiences as an ATC by answering questions 1 through 4 and question 8. Additionally, they were asked to describe the priorities of the athletic department, the role of ATCs in the high school setting, and the challenges that ATCs face in the high school setting.

The interviews ranged in length from approximately 35 to 105 minutes. Eight interviews were conducted by phone, and 7 interviews were conducted in person, based on feasibility and availability. Participants gave advanced written and verbal consent to tape record the interviews. The tape-recorded interviews were then transcribed and analyzed using a grounded theory approach. Data were collected until theoretic saturation was achieved.[15]

The grounded theory approach, as discussed by Glaser and Strauss,[16] is helpful for generating theory (a set of explanatory concepts) based on the data collected. I specifically used open, axial, and selective coding procedures documented by Strauss and Corbin.[15] Raw data were analyzed inductively, and incidents or experiences related to the phenomenon under investigation were identified and labeled as a particular concept. This type of coding strategy is described as "creating tags," and the purpose is to produce a set of concepts that represents the information obtained in the interview.[17] Identifying these concepts and placing them into like categories based on their content completed the formal open-coding process. Relating categories with any subcategories that might exist and examining how one category related to another completed the axial-coding process. Selective coding involved integrating the categories into a larger theoretic scheme and organizing the categories around a central explanatory concept, specifically, the proposition that informal learning processes were critical to the successful professional socialization of ATCs in the high school setting.

Trustworthiness

Several techniques were employed in order to establish trustworthiness of the data collection and analysis, including peer debriefing, data-source triangulation, and member checks. A peer debriefing was completed by having an athletic trainer with a formal education in qualitative methods (a minimum of 3 qualitative research methodology courses at the doctoral level) review the documented concepts and thematic categories for relevance, consistency, and logic. Moreover, the reviewer examined the interview questions in each transcript to determine if they were "leading" in nature. The textual data from any questions identified as being leading were not included in the analysis. The reviewer was in agreement with the findings based on the purpose of the study and even suggested other concepts that would strengthen one particular category.

Data-source triangulation, which is cross-checking perspectives,[18] was obtained by interviewing current high school athletic directors and a former high school ATC. Member checks were completed electronically by e-mailing the results to 5 participants and allowing them to comment on the thematic categories. Three individuals responded, agreed with the results, and had no further input, indicating no misinterpretation of the professional-socialization process that emerged from this study. On an informal basis, I also explained the results to 4 other participants, and they were in agreement with the findings.

Results

The concepts identified during the open coding were organized into 2 categories that gave insight into the professional socialization process: (1) an informal induction process, aspects of organizational learning, and (2) creating networks for learning. The Figure provides a conceptual framework and identifies representative sample concepts and the categories into which they were placed. The axial coding allowed for making connections among categories and understanding the setting that influenced the socialization process. The selective-coding process allowed for the discovery of an overarching theme or informal learning processes that integrated the categories and gave insight into the resultant proposition that informal learning processes are critical to the successful professional socialization of ATCs in the high school setting.

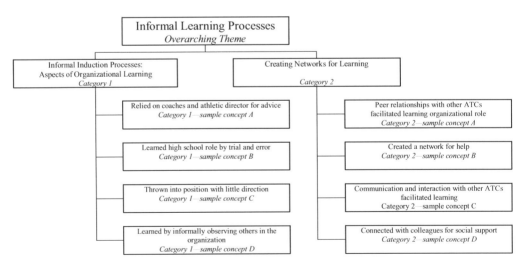

A conceptual framework of the qualitative data organization is presented. Note that the concepts organized into the categories are representative samples of the concepts coded in the margins of the transcripts.

An Informal Induction Process: Aspects of Organizational Learning

Certified athletic trainers entering the high school level reported extremely nonstructured, informal processes relative to learning the full extent of their professional responsibilities within the organization. Peer relationships drove most of the learning at the high school level in that ATCs were able to gain an understanding of their responsibilities by obtaining feedback from the coaches that they worked with and, although in some instances to a lesser extent, their athletic directors (Figure, sample concept A, category 1). For example, Theo stated that learning the "ins" and "outs" of the organization came from interactions with the coaching staff, specifically the football coach:

> He's an extremely intelligent human being and he knows the politics of the system and he knows. . . everything about the "ins" and the "outs" of the school district, and schools in general, so he was very helpful in getting me to learn more about the after-school athletic type aspect [of my job].

Supporting this claim, Alicia stated that "the football coach kind of showed me the ropes. . . ," indicating that the positive relations with the coaching staff influenced her organizational learning when she initially entered the high school setting.

Based on the interview data, the peer relationships may have allowed for informal learning because the organizations were focused on academics and on the health and welfare of the students. That is, during the interview process, there were occasional discussions regarding minor conflicts between staff members, but absent from the interviews were identifiable power struggles that deterred high school ATCs from achieving their professional mission of providing quality health care. In fact, there was evidence to suggest that a common priority at each of the organizations allowed ATCs, coaches, and athletic directors to coexist in a collegial manner. For example, when asked about the organizational priorities, Theo stated:

> Sports. . . definitely. . . helps define the high school. So I think the administration accepts that. . . [but] definitely the school's number one mission is the education of its students. Our athletic department is very strict on its maintaining of the academic requirements for eligibility. . . we actually started an at-risk after school program for any kid who is in danger of failing a class. . . They get help with any questions they may have and they are required to do their homework before they go to practice or any game. I firmly believe I have one of the best coaching staffs to be able to work for as far as the kids come first and the victories don't.

Both athletic directors interviewed for this project supported the ATC's comments:

> Well, the priority certainly. . . [is] the student-athletes, students first. So academics is very important to us here at High School A. We watch and monitor that very closely and all our coaches have that mindset. Then, certainly, the second would be the safety issue. That comes from my athletic experience, prevention. We are very attuned to proper weight training and nutrition. We constantly talk to student-athletes about diet, sleep, healthy choices for their bodies, [both] mentally and phsyically. Then, we feel that winning will take care of itself.

Academics are first and foremost in my mind. And, I say athletics is an educational opportunity and so what we try to do is make. . . our mission [reflect this priority]. I make all the decisions based on that particular mission.

Participants in this study were socialized into their role in the high school setting using individual and informal tactics, meaning that they were often functioning independently and relying on trial-and-error learning (Figure, sample concept B, category I). Moreover, many participants explained that they learned through the observation of others (Figure, sample concept D, category l). For example, Marsha stated, "I am very fortunate that I've been around a lot of people who have very good people skills, and I feel that I picked up on that. I think athletic trainers are very adaptable to being flexible and can learn some of the ropes just by watching."

Additionally, with the exception of working closely with coaches and other staff, there was no formalized mentoring in place to facilitate learning of their role. In short, the ATCs were expected to enter the setting and immediately begin functioning in their role in an independent manner (Figure, sample concept C, category I).

There appeared to be no detectable timeline to the professional-socialization process relative to specific events or professional experiences, except that when each of the individuals entered the setting, there was a small period of adjustment. The induction process during this time was extremely informal. For the high school ATCs who obtained a position immediately after completing their undergraduate degree (Carol, Theo, Alicia, and Payton), much of the adjustment was related to moving from a setting that had a great deal of supervision to one with little supervision. For example, Theo mentioned:

[Although] the first year was pretty calm for the most part, [I] just tried to get used to things and being on my own without having a supervisor that could help me out with whatever I needed help with.

Even Reginald, reflecting on his first-year experience, stated, "I knew what I was doing [relative to] athletic training. . . but I didn't know all the 'ins and outs' and the politics."

Consistently, participants identified the necessity of setting up their health care system in the high school setting and learning how best to communicate with other staff (coaches and athletic directors), parents, and student assistants and how to meet the demands of providing health care to such a large athletic population. Learning this role was informal and often described as a trial-and-error process. For example, Alicia reflected on initially entering the high school setting and stated:

Actually, I learned by trial and error. There really wasn't anyone [to facilitate my learning]. The other person that I worked with started the same day I did, so neither of us had any idea, we just kind of went into it with what knowledge we had of what a college [athletic] training room was like and kind of took that into our own [facilities]. Obviously we had to modify certain things, like getting prescriptions for modalities and things like that.

Similarly, when asked about how she initially learned her role in the high school, Jennifer commented on the induction period and stated that "unless there was previously an athletic trainer within the organization, there would be nobody to orient a newcomer to [his or her] role." Confirming this theme, Robert stated, "I was not given any type of orientation or didn't meet any of the coaches. I was given a set of keys and told 'good luck.' I was on my own. . . My budget was woeful. . . so it was tough that first year. . . " Donald added, "I think a lot of [learning of the job] is trial and error. There are a lot of things that I don't do now that I did years ago and I thought, well, that didn't quite work, I could have handled that situation differently."

In order to successfully navigate the informal induction period and evolve as an ATC at the high school setting, participants sought to create learning networks beyond their institution. The next theme further explains this phenomenon.

Creating Networks for Learning

Participants consistently discussed contacting other ATCs in order to learn how best to deal with their responsibilities. While many ATCs who were new to the high school setting often relied on their previous mentors for advice and direction, it was interesting to find that both novices and veterans at the high school level took the initiative to make connections with other ATCs outside the high school in order to continually learn (Figure, sample concepts A and C, category 2). In fact, Theo stated that while he did not have to travel to away events with the teams, he did so in order to network with the other ATCs at local schools and learn from them:

I have done a lot of traveling with my sports which I don't have to do but because of the bonds I've made with the kids and parents I. . . go to the away games. [I also] get to talk to the other athletic trainers and [discuss various situations] and sometimes they have some helpful answers or places to look for better answers.

Jeremy also identified networking as critical to learning:

I think that is one of the biggest things you learn as you go is that you do develop some type of network of athletic trainers who you can call and say 'this is the [most interesting] thing. I've been doing this [procedure] and this [technique], but [the athlete] is just not getting better. What do you think?'

It appears that the role of a high school ATC evolves as he or she gains more experience. Initially ATCs make network connections in order to learn, but as they become more experienced, they play more of a mentoring role, being contacted by less experienced peers for advice about how to deal with issues in the high school setting. For example, Robert stated:

Professionally. . . my involvement with [the regional professional association] has been really rewarding. It has helped me a lot to deal with life. . . I will often call up Bruce Johnson (pseudonym) and bounce [ideas] off of him. But [now] more people call me and bounce things off of me more than I call other people. I really feel like I'm the grandfather in the area.

Now, I get phone calls all the time from other athletic trainers about 'how should I handle this. . . ?' Maybe it's because I'm old now too, but there is a bigger network now than there was 20 something years ago.

Aside from facilitating learning, networks were also used to provide help and social support (concepts B and D, category 2). For example, Reginald explained that he received a call from a high school ATC in a nearby suburb. The ATC had experienced the death of an athlete approximately 6 months earlier, and called to get contact information for another area ATC who experienced an athlete's death more recently because he wanted to offer his support.

Given the nature of the 2 thematic categories, the resultant proposition is that informal learning procedures are critical to the professional-socialization process for ATCs working in the high school setting. The results of this study suggest that learning through informal means, such as collegial networks, organizational peers, and trial and error, are necessary elements for navigating the high school work setting and being socialized into the ATC role.

Discussion

Professional socialization is a complex process that has the propensity to influence an individual's success in a work environment. The socialization literature[6,10,19-22] concludes that the initial entry into an organizational setting is a period of adjustment for many professionals. This study supports these findings, as participants suggested that there was an initial adjustment when entering the high school setting from their previous setting (undergraduate program, master's program, or previous job). Additionally, many participants learned through informal means such as trial and error and by observing others in the organizational setting. Based on a socialization study by Ostroff and Kozlowski,23 this is not unusual, as many individuals frequently rely primarily on observation of others and trial and error to acquire their information in an organization. Thus, informal learning plays a critical role in the professional-socialization process.

Informal Learning

Informal learning can be defined as a lifelong process by which individuals acquire and accumulate knowledge, skills, attitudes, and insights from daily experiences and exposure to an environment and individuals rather than from a structured, hierarchic education system.[24] Informal learning generally occurs as a means of achieving particular individual or organizational goals, often as a result of expanded responsibilities.[25] Informal learning is reported to account for up to 90% of new learning.[26]

Informal learning is often intentional but lacks the formalized structure that is found in education-based systems.[27] Although some authors[27] contended that trial-and-error learning is better defined as incidental learning because it is not intentional, the participants in this study appeared to make intentional, conscious efforts to identify which professional actions or strategies worked and which did not, in order to better meet the demands of their work environments. Informal learning implies that much of the learning was experiential and action based in nature. That is, learning occurred as participants attempted to find solutions to problems or events related to their work setting.

As the participants of this study entered the high school setting and accepted the organizational challenges involved in developing a system of health care, they engaged in informal learning to manage their new responsibilities. Leslie et al[25] stated that more informal learning takes place when the goals of an individual and an organization are in alignment. As in this study, a common priority appeared to be shared by ATCs, athletic directors, and coaching staffs that suggested the student-athletes' welfare related to academics and health was a main concern. Perhaps this allowed for a great deal of informal learning to take place.

Garrick[28] stated that many respected adult educators link informal learning to concepts such as "autonomous learning," "self-directed learning," and "independent learning." In fact, supporting these claims, Lankard[29] suggested that to enhance informal learning, learners must (1) autonomously direct their learning, (2) self-evaluate their learning, (3) engage in critical self-reflection, and (4) think critically. This has many implications for athletic training preservice and continuing education.

Implications for Preservice Athletic Training Education

Athletic training preservice education has thoroughly emphasized the necessity of addressing content, competence, and clinical proficiency. This study, however, identifies the necessity of fostering reflective practitioners who are self-directed and self-evaluative to fully prepare them as informal learners. Educators can enhance these aspects by using reflective journals, individualized learning plans, and formalized student self-evaluations. Although these characteristics may be approached in some programs and curriculums, given their importance to professional development, consideration must be given to making these goals explicit during undergraduate athletic training education.

Implications for Continuing Athletic Training Education

The previously mentioned characteristics have implications relative to continuing professional education as well. For example, Bickham[30] suggested that the "foremost goal of continuing professional education is to teach professionals to develop and hone critical-thinking skills." Continuing professional educators who give ATCs opportunities to conduct verbal reasoning and problem solving and who consciously raise questions can help accomplish this goal.[30] Unfortunately, a great deal of continuing education in athletic training focuses on content, such as transferring information by listening to lectures, completing home study courses, or updating cardiopulmonary resuscitation skills, rather than focusing on fostering critical-thinking ability. Moreover, although the transfer of information at a continuing education event can be enlightening for many, a significant issue is whether the information improves professional practice.[31]

Adult learners will engage in learning activities providing their job performance will be enhanced by the experience.[32] Given the participants' need to learn information in order to solve problems linked to practice, continuing education must be more integrated with practice-based problems if it is to be effective.[30] Perhaps this is why the participants in this study created learning networks. Ritchie[33] commented on networking and stated that "while professionals tend to be self-directing and autonomous, they are not necessarily singular or practicing in isolation. Professionals rely on other professionals to help meet their continuing learning needs."

Given the self-directed nature of informal learning, perhaps alternative strategies to continuing education, such as self-directed learning plans based on an individual's contextual learning needs are a viable continuing education strategy.[34] Such a strategy would foster self-direction, self-evaluation, and self-reflection.[33] Additionally, critical self-reflection strategies, such as video analysis of skills and self-evaluations, can promote intentional active learning and the formation of learning communities that facilitate practice-based learning[33] and support the direct link of applying continuing professional education to practice.

Limitations

Most of the participants in this study were practicing at schools located in metropolitan areas as designated by the US Department of Commerce, Bureau of the Census, yet none were located in an inner city school or extremely rural setting. The propositions resulting from this grounded theory, therefore, may not be transferable to the inner city or rural school context.

Conclusions

The purpose of this study was to gain insight into and understanding of how ATCs in the high school setting initially learned and continued to learn their professional responsibilities in an organizational setting. The organizational aspects of the professional-socialization process among high school ATCs are principally informal in nature. As such, the ATCs relied on informal learning strategies during their period of induction. To facilitate their continued development, informal learning networks that largely comprised colleagues outside of the organizational setting were created.

To ensure that individuals effectively learn through informal means, preservice athletic training education programs would be well advised to foster the development of reflective practitioners who think critically and are self-directed and self-evaluative. This can be accomplished by using such educational strategies as reflective journals, learning plans, and independent projects.

Continuing professional educators should also attempt to foster self-evaluation, critical reflection, and critical-thinking ability. Continuing professional educators can accomplish this by employing strategies such as verbal reasoning and problem solving and consciously raising questions and giving clinicians an opportunity to discuss their thought processes in a nonthreatening learning environment. Moreover, it has been argued that continuing education should be linked to practical problems.

Because informal learning is highly contextual, multiple influences have the propensity to shape the extent to which informal learning is successful, including cultural factors, career structure, technology, and learning needs.[35] As such, future studies could investigate exactly how these factors influence informal learning and role socialization. Additionally, because informal learning can be inhibited in many ways, it may be helpful to examine the environmental inhibitors (ie, job demands) of informal learning in various athletic training settings.

Acknowledgments

This study was funded by a grant from the Mid-America Athletic Trainers' Association.

References

1. Clark PG. Values in health care professional socialization: implications for geriatric education in interdisciplinary teamwork. *Gerontologist.* 1997;37:441-451.

2. Harvill LM. Anticipatory socialization of medical students. *J Med Ethic.* 1981;56;431-433.

3. Hayden J. Professional socialization and health education preparation. *J Health Educ.* 1995;26:271-276.

4. Becker HS, Geer B, Hughes EC, Strauss AL. *Boys in White.* Chicago, IL: University of Chicago; 1961.

5. Shuval JT, Adler I. The role of models in professional socialization. *Soc Sci Med.* 1980:14:5-14.

6. Glen S, Waddington K. Role transition from staff nurse to clinical nurse specialist: a case study. *J Clin Nurs.* 1998;7:283-290.

7. Cohen HA. *The Nurse's Quest for a Professional Identity.* Menlo Park, CA: Addison-Wesley; 1981.

8. Lyons M. Understanding professional behavior: experiences of occupational therapy students in mental health settings. *Am. J Occup Ther.* 1997; 51:686-692.

9. Corb DF, Pinkston D, Harden RS, O'Sullivan P, Fecteau L. Changes in students' perceptions of the professional role. *Phys Ther.* 1987;67:226-233.

10. Teirney WG. Rhodes RA. *Faculty Socialization as a Cultural Process: A Mirror of Institutional Commitment.* Washington, DC: George Washington University, School of Education and Human Development; 1993. ASHE-ERIC Higher Education Report No. 93-6.

11. Pitney WA, Parker J. Qualitative research in athletic training: priniciples, possibilities, and promises. *J Athl Train.* 2001;36:185-189.

12. Merriam SB. *Case Study Research in Education: A Qualitative Approach.* San Francisco, CA: Jossey Bass; 1988.

13. Maxwell JA. *Qualitative Research Design: An Interactive Approach.* Thousand Oaks, CA: Sage Publications Inc; 1996.

14. Chenitz WC, Swanson M. eds. *From Practice to Grounded Theory: Qualitative Research in Nursing.* Menlo Park, CA: Addison-Wesley; 1986.

15. Strauss A, Corbin J. *Basics of Qualitative Research: Techniques and Procedures for Developing Grounded Theory.* 2nd ed. Thousand Oakds, CA: Sage Publications Inc; 1998.

16. Glaser BG, Strauss AL. *The Discovery of Grounded Theory.* Chicago, IL: Aldine; 1967.

17. Côté J, Salmela JH, Baria A, Russell SJ. Organizing and interpreting unstructured qualitative data. *Sport Psychol.* 1997;7:127-137.

18. Patton M. *Qualitative Evaluation and Research Methods.* 2nd ed. Newbury Park, CA: Sage Publications Inc; 1990.

19. Colucciello ML. Socialization into nursing: a developmental approach. *Nursing Connections.* 1990;3:17-27.

20. Stroot SA, Faucette N, Schwager S. In the beginning : the induction of the physical educator. *J Teach Phys Educ.* 1993;12:375-385.

21. Williamson KM. A qualitative study on the socialization of beginning physical education teacher educators. *Res Q Exerc Sport.* 1993;64:188-201.

22. Flynn SP, Hekelman FP. Reality shock: a case study in the socialization of new residents. *Fam Med.* 1993;25:633-636.

23. Ostroff C, Kozlowski SWJ. Organizational socialization as a learning process: the role of information acquisition. *Personnel Psychol.* 1992;45:849-874.

24. La Belle TJ. Formal, nonformal and informal education: a holistic perspective on lifelong learning. *Int Rev Educ.* 1982;28:159-175.

25. Leslie BH, Kosmahl-Aring M, Brand B. Informal learning: the new frontier of employee & organizational development. *Econ Dev Rev.* 1998;15: 2-18.

26. Lohman MC. Environmental inhibitors to informal learning in the workplace: a case study of public school teachers. *Adult Edu Q.* 2000;50:83-102.

27. Marsick VJ, Watkins KE. Informal and incidental learning. *New Direct Adult Contin Educ.* 2001;89:25-34.

28. Garrick J. Informal learning in corporate workplaces. *Hum Resour Dev Q.* 1998;9:129-144.

29. Lankard BA. *New Ways of Learning in the Workplace.* Columbus, OH: ERIC Clearinghouse for Adult. Career, and Vocational Education; 1995. ERIC Digest #161.

30. Bickham A. The infusion/utilization of critical thinking skills in professional practice. In: Young WH, ed. *Continuing Professional Education in Transition: Visions for the Professions and New Strategies for Lifelong Learning.* Malabar, FL: Krieger; 1998.

31. Cevero RM. Trends and issues in continuing professional education. In: Mott VW, Daley BJ, eds. *Charting a Course for Continuing Professional Education: Reframing Professional Practice.* San Francisco, CA: Jossey-Bass; 2000.

32. Knowles M. *The Adult Learner: A Neglected Species.* Houston, TX: Gulf Publishing; 1990.

33. Ritchie K. Informal, practice-based learning for professionals: a changing orientation for legitimate continuing professional education? *Aust J Adult Commun Educ.* 1998;38:69-75.

34. Pitney WA. Continuing education in athletic training: an alternative approach based on adult learning theory. *J Athl Train.* 1998;33:72-76.

35. Hager P. Recognition of informal learning: challenges and issues. *J Vocation Educ Train.* 1998;50:521-535.

Appendix B

Darwinism in the Gym

Clive C. Pope, The University of Waikato

Mary O'Sullivan, The Ohio State University

Clive Pope is with the Dept. of Sport & Leisure Studies, The University of Waikato, PO Box 3105, Hamilton, New Zealand; Mary O'Sullivan is with the College of Education, The Ohio State University, 149A Arps Hall, N. High St., Columbus, OH 43210.

This study examined the ecology of "free gym" as it occurred in both school lunch hour and after-school community settings. In an effort to understand how urban youth experiences sport, an ethnography using multiple methods was conducted to ascertain how urban youth shape their own cultures according to the social forces operating within the gymnasium. A period of sustained observation revealed a student-imposed hierarchy that was dominated by skilled male African American basketball players. Status was gained through what occurred within the free-gym ecology. Students often had to learn the system by "serving time" before they could join a desired level of the hierarchy. While a few students thrived in this environment, most merely survived or were marginalized. Such a setting has implications for how physical education and school culture is subjected to wider societal influences. The presence of socially chronic situations such as free gym require a pedagogy that is more democratic and more enriching, thereby moving from the real toward the ideal.

Key words: extracurriculum, youth culture, gymnasium ecology

…the socio-cultural dynamics that shape our lives in a larger society must be considered and analyzed in the instructional activities within sport if a full, rich, and accurate picture of sport pedagogy is to emerge. (Schempp, 1998, on-line)

Over 25 years have passed since Larry Locke shared his views of how important it is to reveal what goes on in the gymnasium and expose what the tourists never see. He argues for the need to examine the ecology of the gymnasium in order to understand teaching and learning (Locke, 1974). Life in the gym, according to Locke, was marked by complexity, diversity, time constraints, and the special nature of the subject matter of physical education. In particular, visiting the gymnasiums could shed light on the physical education teaching and physical education as a subject.

Locke (1974) warned that much of the action that occurs in the gym is often unrecognized. While regular visitors may become accustomed to the movement and the sounds, "as full time actors in the play of roles that revolves around the gym, they often sense only the consequences of events without really discovering the social machinery that causes things to happen as they do" (p. 4). To better understand the world of physical

Entire text of appendix B is reprinted, by permission, from C.C. Pope and M. O'Sullivan, 2003, "Darwinism in the gym," *Journal of Teaching in Physical Education* 22: 311-327.

education and those who teach it, he called for prolonged attendance to learn about the microcosm of gymnasium life.

However, researchers are not the only visitors to the gymnasium. The long-term visitors are the students, also tourists, who form part of the ecology. Their needs and experiences are often shaped by *how life is* within the walls of the gym. Moreover, what is learned beyond those walls in the form of an urban "street literacy" can seem contradictory to what teachers may seek to promote in educational settings, which often requires cooperation from young people (Cahill, 2000; Katz, 1995).

The purpose of this study was to examine the sport experiences of youth at one urban high school in a large U.S. city during the "open gym" at lunchtime, and during recreation times after school at selected community recreation centers. We endeavored to interpret how those experiences contributed to their physical education or mis-education. The focus of this study was on the sport (formal and informal) experiences of high school youth and not merely high school sport. While the students who participated were based at one high school, their sport experiences went well beyond traditional sport contexts. In addition, the study attempted to ascertain the opportunities and constraints that influenced these urban youth. For many students, their only experiences of sport occurred either in physical education as part of the formal curriculum, or in other programs offered in high school that fall under the heading of the extracurriculum.

Theoretical Framework

The Extra Curriculum

Broadly conceived, education includes teaching and learning, socialization and enculturation, and takes place in formal, informal, and nonformal situations (Hansen, 1979). Like the functional curriculum, extracurricular activities can be touched by wider societal problems, and research reviews have pointed out that extracurricular participation contributes in vital ways to adolescent development (Berk, 1992).

Eder and Parker (1987) highlight the potential impact that schools can have on adolescents' values and behavior through extracurricular activity. They offer three reasons: First, unlike the structured nature of classroom contexts (Cusick, 1973), extracurricular activities hold the potential to become social events marked by interaction and meaningful participation. Second, the social dimension of extracurricular activities holds potential for interaction with the opposite sex. This issue has particular relevance in the U.S., where many physical education classes are sex-segregated. Third, student visibility is elevated amidst the peer group during many extracurricular activities. This enhanced status can in turn intensify membership in the peer culture.

Little is known about how young people consciously experience sport. Moreover, expressions of their needs have seldom been communicated directly to adults. The criteria young people adopt to evaluate their involvement in sporting experiences are often dependent on changing life situations (Brettschneider, 1990). Not enough attention has been allocated to this important component of the youth sport milieu, and therefore little is known about how youth receive, regard, define, and experience sport especially as it relates to informal sport.

A lack of opportunity to play is symptomatic of many U.S. school programs that usually support only one team per sport. While suburban youth may have a variety of sport opportunities, for inner city students the school may be their only opportunity for participation (Eitzen, 1995; Kozol, 1991). Their experiences are often hobbled by access and provision (Ewing, Seefeldt, & Brown, 1996). The presence of a sport delivery structure that should cater to urban youth has resulted in an unfortunate paradox: "the general response to a growing concern about at-risk youth in the 1990s is to make school sport less accessible through the institution of pay-to-play plans" (Siedentop, 1996, p. 272). The result is a decline in youth sport among urban adolescents.

Decreased opportunities for many youth appear to have promoted two related trends. The first trend, fueled by their continued interest in sport, is the shift on the part of youth to more passive forms of sport participation. Because they are unable to be players, many, like their adult counterparts, subscribe to sport in the more passive role of spectator. The second trend, supported by increasing youth choices, has been a growing conflict of interest between academic pursuits, part-time work, and indulgence in at-risk behavior as alternatives to the demands of many sport programs.

Revisiting the Gym

To learn more about the educational value of high school sport, we need to examine youth and sport at a microanalytic level. Grupe and Kruger (1994) argue that there is a need to "be concerned with the discussion of values, ethics, and moral problems of sport and sport education" (p. 20). To enter into such discussion we need to examine the role of sport in the current value system of today's youth. Moreover, it is prudent to adopt an ecological framework (Bronfenbrenner, 1979, 1989) to examine what norms and values about sport are transmitted to today's youth by significant others.

High school students' participation in active leisure forms (such as sport) is determined largely by what they are interested in, which in turn is influenced by the availability of suitable resources (Garton & Pratt, 1987, 1994). The issue at stake is the degree to which such responses are suitable for urban youth. Furthermore, the dominance of sport at the high school level is seen by the sustained salience of sport within the value structure of many urban youth. Young people gain acceptance by excelling in something that is valued by their peers (Evans & Roberts, 1987).

The cloistering of students in a social cauldron like the gymnasium has promoted the peer culture (Brown & Theobald, 1998). Such a culture can affect the relationships that are formed and the activities that are selected in a given context. As relationships form, individuals associate with like individuals and establish groups or crowds. The selection process may not always be at the discretion of the individual; rather his or her affiliation may be determined through group or crowd selection.

The goal of this study was to examine what occurs in urban school and community gymnasia during the time known as "free gym." A secondary goal was to examine the social interactions in those spaces and the implications for physical education, and for those who are taught.

Berk (1992) reports that a "comprehensive review of the empirical literature...yielded few ethnographic accounts...aimed at discerning the quality of students' experiences...in extracurricular activities" (p. 1036). Berk argued that this was due to a "preoccupation of educational researchers with the quality of the American formal academic program" (p. 1035). To understand more about the ecology of physical education contexts, we must examine the nature and impact of other contexts because, as Brown and Theobald state,

> Unresolved issues still fester, however, particularly concerning the major objectives that should underlie an extracurricular program, the nature and degree of

connection that should exist between extracurricular participation and classroom learning, and the degree to which schools should take ownership of these activities. (1998, pp.116-117)

This study involved revisiting the gymnasium and learning more about the social machinery that has an impact on youths' access to sport and the nature of urban youths' sport experiences.

Methods

Shady Woods is an urban high school in a large American city. It has an enrollment of over 800 students, of whom 66% were on a free or subsidized-lunch scheme. In all, 65% of Shady Woods students were identified as African American, 25% as Caucasian, 3% as Asian American, while the remainder included Hispanic, Native American, or "other" ethnic groups. Boys slightly outnumbered girls (51 vs. 49%). Observation was the research tool by which we could gain an understating of the setting. It allowed us to note the events and behaviors of the major players. Ethnographic research can rely on varying degrees of participant observation. The practice of observation was largely conducted using the more passive or peripheral forms of complete observer or participant as observer. It was considered beneficial to switch between these two derivatives to gain a sound understanding of local ecology without disrupting the existing social machinery at each of the three sites. Multiple methods were used to gain a better understanding of the immediate social circle and the wider cultural environments of these urban youth. The aim of participant observation was "a commitment to adopt the perspective of those studied by sharing in their day-to-day experiences" (Denzin, 1989, p. 156).

The works of Spradley (1980) and Werner and Schoepfle (1987) assert that observation can pass through a three-phase process of "descriptive observation" which leads to "focused observation." After attending to what is considered more important, observation may then proceed to the final phase of "selective attention" whereupon specific features of chosen activities are studied. Observation was selected to gain an appreciation of free gym because "ethnography appears a particularly well suited endeavor for the sport pedagogy scholar in search of the forces of social change that reside within the teaching/learning process of sport" (Schempp, 1998, on-line).

Lunch-Hour Observation

We visited free-gym sessions at the Shady Woods High School gymnasium over a period of 5 months. During each visit we maintained a position on the top row of the bleachers or at the edge of the basketball court. The former strategy was to gain a bird's eye view of the dynamics and behavior of whoever was on the floor; the latter strategy focused on particular groups or on the interactions between students. The free-time sessions occurred between 10:30 and 11:30 every morning. A teacher (usually a physical educator) allocated five of six basketballs and then positioned him/herself at the doorway of the gym. Student attendance varied depending on who was playing, the weather, or events elsewhere in the building that day. On average, between 80 and 130 students occupied the gym. The numbers would fluctuate during the hour and would usually increase toward the end of lunchtime. Occasionally one of us would talk with a student or a staff member who was present, but generally our role was purely that of observer.

Community Observation

From early conversations with Shady Woods students, two recreation centers were identified as sites where students would "hang" after school. One center, Northside, was three blocks from Shady Woods while the other, Crosstown, was situated several miles across the city. "Northside Recreation Center" was visited four times while "Crosstown Recreation Center" was visited twice. Trips to these two venues were mostly straight after school or in the evenings and each visit would last 90 to 120 minutes. The number of Shady Woods students in attendance at these two centers fluctuated between 5 and 40.

Fieldnotes

At each of the three venues fieldnotes were taken. This would involve either talking into a microcassette recorder or writing in a small notebook, depending on the context. If the observer's presence was unobtrusive, a cassette recorder was used because it produced a better record when there was a lot going on. If the observer was close to other people or if it was not possible to use a cassette to record observations, these were entered into a notebook at the earliest convenience. The observer also made comments into the tape recorder upon returning to the car or while driving between venues.

Observation entailed constant scanning from the periphery of a venue and "observing out loud" (Allison, 1988). It was a matter of focusing on quite general things at first before familiarity allowed a more concentrated attention. Initial observation would ask: Who was there? What age and gender were they? What were they wearing? Who were they interacting with? What were they doing? How long did they do it for? How were they arranged? When possible, the observer would revisit sites with new and more focused questions. However, students who visited the venues varied significantly each day and this often restricted how questions could be answered.

The final component of the observation process involved making entries into a log on a home computer. This log became a repository for ideas, events, hunches, and issues. Each page of the log was divided into three columns: observational notes, personal notes, and theoretical notes. Entries made in the log would be reread regularly before returning to the observation sites.

We both adopted strategies to address trustworthiness during the data collection and analysis process. First, a peer debriefing strategy ensured that it would be the respondents' rather than the researchers' categories that would dominate the findings (Lincoln & Guba, 1985). This included regular meetings during which findings and interpretations were audited to establish a consensual validation. Moreover, we reworked the assertions until we felt we were representing the students' thoughts and actions and not imposing our own viewpoint. The final strategy involved prolonged involvement with those being studied, gathering from multiple data sources (interviews and observations) of the students in the environment.

The analysis assumed a two-tiered approach: a descriptive review of the contextual characteristics that influenced free gym; and an interpretation of the data that supported the task of asking why specific results had emerged through observation. Each characteristic was examined inductively to establish lower level descriptors that contributed to higher lever categories (Patton, 2001). This process was constantly adopted to confirm, modify, disregard, or merge raw data into dominant themes and categories.

Results and Discussion

A major finding was that a student-imposed hierarchy determined the nature and degree of participation each student could assume during free-gym play. Any attempt by students to elevate their status within the free-gym ecology required that they do their time by serving an apprenticeship at a subservient level. In addition, a degree of movement literacy was expected of practicing and potential players as well as observers. To survive or thrive in this ecology, a participant has to possess a certain degree of street literacy involving both physical ability and "street smarts."

If the meanings that many youth develop about sport are based on personal experience, then what occurs during free-gym sessions at the school and community facilities may discourage participation among a large number of youth. Informal and unstructured games, minimal or no adult presence, and casual groups of spectators who seldom stayed for long periods of time marked the two field settings. Virtually every site was marked by pick-up basketball games, a strong social ambiance, and a clear but implicit set of rules for behavior. It was a place to hang out with friends and to check who was playing or waiting to play.

Lunch-Hour Free Gym: A Cultural Caldron

The Shady Woods gym is a double size basketball court and, with the bleachers retracted, allows students to utilize all six basketball rings. There was usually one basketball available for each basket, and the lunchtime bell signaled a race to get one of the balls and ensure a chance to play at one of the baskets. This venue was a place to gather, to greet friends, to hang out, and to compete. Every morning between 10:30 and 11:30 the gym was open to all students. A staff member, usually a physical education teacher, was stationed at the only door through which students could enter or exit. During this hour the gym was frequented by large numbers of students.

The only activity pursued was basketball, or what the young people referred to as "ballin." The activity was spontaneous, unstable, intense, and very public. Once a student found a ball, he or she would congregate with peers and organize a pick-up game. It was usually a small-team game such as 3 v 3 or 4 v 4. Each contest was characterized by long periods of verbal jousting or "trashing" while some player walked around and around the periphery of the half court area, waiting for an opportune time to begin play. The other version of "ballin" was taking turns at shooting from self-selected sites, with the shooter keeping the ball if he was successful. The shooting version was very unstable as the activity could break up at any time or the number of participants could change rapidly.

On most days there were over 40 students on the floor 5 minutes after the bell rang to begin the lunch period. As the lunch hour continued, the numbers swelled, depending on the day of the week and what else was happening in the school. On the floor 90% of the participants were boys and mostly African American. When girls played, they always played at the basket nearest the door; this was the least favored venue as students walking in and out of the gym often interrupted games at that basket. On several days that same basket would be taken over by boys, or by some girls who stood their ground or who compromised and allowed a mixed-gender game.

The geography of this venue altered according to who was present. A power structure was obvious. Certain students determined who played at what baskets and how much space they had to play their game. If certain male athletes decided they wanted to play, they would often take one of the two small courts to play a game in front of a reasonable size crowd. Occasionally they would set up a game on the main court, forcing other players off the floor or confining them to a very small area under the four side baskets. These players were often juniors or seniors on the school basketball team. If they began a game, it would attract several spectators. On occasion some students refused to give up their game space on the side baskets and endured players running through their area.

The young people who frequented the gymnasium fell into loosely defined groups somewhat akin to the categories of students Griffin (1984, 1985) observed in the middle school physical activity setting. Altogether there were 7 groups:

1. *Bullies.* Bullies were always male, usually senior, and often skilled performers. They often dictated what game could be played and where. Their games usually resembled an exhibition in slam-dunking with virtually no midcourt action. The more spectators present, the more intensely the Bullies performed. As the intensity of their games increased, they would spend more and more time dribbling the ball at the top of the "key" and just verbally trash (ridicule or abuse) each other, exchanging jibes for several minutes before beginning a new play for the basket. Their most intense performances were on days when the basketball team members were not involved in interscholastic competition. If a Bully was not playing, he often positioned himself alongside other peers (some of whom were Bullies) beside another game and trashed the players on court. This behavior was particularly popular as Bullies often fed off each other as they trashed their selective targets:

> A tall, lean African American male calls from the sidelines. "You want some of this?" One of the players replies, "We'll take the money," at which point he launches a pass like a bullet to a teammate who shoots only to have it rejected by an opponent… The sideline observer replies "sad," which draws an icy stare from the unsuccessful shooter who has retrieved the ball and duly heaves it toward the roof, narrowly missing a suspended light. (Fieldnotes)

2. *Jousters.* Jousters consisted of a small group of boys intent on disrupting as many games as they wished to risk. They would often steal a basketball and run to another basket with it. They often did this to girls who might be playing a game. In addition, Jousters would intrude during a game and often try to slap the ball out of a player's hands:

The Court 4 game is interrupted by an invader who tries to slap the ball from its possessor unsuccessfully. Having failed to gain possession of the ball, the intruder then attempts to foot-trip his elusive target. Again, he is unsuccessful and proceeds to move onto Court 3. (Fieldnotes)

> His action is unnoticed by the duty staff member who is positioned a significant distance away.

> Another behavior was to stand right under the basket in the middle of a game, thereby forcing some form of confrontation or retaliation. Some Jousters enjoyed selecting a player and just punching him a few times or kicking at him to get a response. Some would use these tactics to get into a game and, if they didn't succeed, they would move around the baskets like nomadic disturbances. This self-selection

process was quite contrary to the subtle negotiation that took place between other participants. Invariably Jousters were not skilled basketball players and would not have been picked for games.

3. *Posers.* Posers were some 10 boys who seldom participated but used this forum to show peers their physique and what they could do. A favorite activity of several Posers who worked out in the weight room was to strip to the waist and walk around the gym with their top draped around the waist. Their upper body would often be adorned with gold chains. Some of this group would perform pull-ups on the basketball ring during games. Other behaviors included punching the mats on the walls vigorously several times or kicking any ball that strayed in their direction. Occasionally a Poser would try to get the attention of people on the floor. It was not uncommon for a Poser to put on a pair of red sweat pants, roll up one leg, and walk around the floor. Demonstrations of this nature were generally ignored by other free-gym participants and the teacher supervisors.

4. *Benchies.* Benchies included some girls but were usually boys. On most days there were 30 to 40 students in this group. They would hope to play but often spent the entire lunch hour just waiting for a chance to play. They usually congregated directly under the basket to make sure the players were aware of their intentions. Benchies would wait to be invited on court, or they drifted onto the court when pick-up games were starting and hope to be included in the numbers.

A junior student leans against the wall behind the basket of Court 5. He wears a Nuggets 55 top and has long silky shorts that hang to his kneecaps. He constantly monitors the periphery of the court sizing up other wannabe players. As a game breaks up, he is one of several students to meander onto the floor and assemble under the basket while a few shots are thrown up. With one eye cast toward the basket, he is also constantly panning the court to catch the attention or eye contact of his peers. Some of the players have collected in small groups that appear to be the next game teams. His presence is not required and, once again, he resumes his position leaning on the wall to continue his vigilant role. (Fieldnotes)

> Benchies would sometimes have friends or classmates who were on court; they usually arrived early and changed into their basketball hightops or stood holding them in case they might be called onto the court.

5. *Hangers.* Hangers were evenly made up of boys and girls who would turn up in the gym just to see what was going on and who was there. They often congregated by the doors with the supervising teacher, drifting in and out during lunch hour. They would bring food or drink into the gym and share with friends during their conversations. This gesture was in direct violation of school rules but there seemed to be a total disregard for such a violation. Hangers usually frequented the gym if there was not much happening elsewhere in the school. If a good game was occurring, they would somehow get the message and enter en masse. A confrontation between players would also ensure the arrival of the Hangers to the scene. Some group members merely used the gym as a place to chat or read a book in the company of others.

6. *Venerators.* Venerators were almost all well-dressed girls, often in the 9th or 10th grade. For this group the gym was a place to check out other people as well as be seen by significant others, particularly male athletes. Venerators made their presence felt

by walking up and down the sideline or selecting a performer and trying to catch his attention. The more confident Venerators would often try to start a conversation with a player, and if that failed, they would begin baiting him with jibes or quips.

Josh, a member of the varsity team, receives the attention of three admirers. A successful 3-point shot draws the response, "You good," which Josh tries to ignore but is unable to conceal a smirk on his face. His three admirers giggle and push each other, causing Jenny, who is dressed in high heels, to stumble and look very embarrassed. Jenny asks, "You teach me how to do that?" to which Josh retains his grin but no further response occurs. He walks away from them and his departure earns a response, "Be cool," from Jenny, who has now regained her balance and her composure. She pretends to ignore him and convinces her two friends that they are wasting their time. (Fieldnotes)

The girls worked in groups of two to four. If the selected target did not acknowledge their attention, they would often get louder and more conspicuous in their behavior until they got his attention.

7. *Contestants.* Contestants were those who played basketball. With the exception of three or four girls, this group was all male and mostly African American. They ignored almost all around them and became engrossed in their pick-up games. These participants were skillful players. They took their competition seriously and played with intensity. Contestants played by agreed-upon rules and usually remained on court until the bell rang. They were often quite vocal both on and off the court. Most of them were confident and extremely competitive. Although they usually played in 4 v 4 games, they often indulged in 1 v 1 plays within the game. Their goal each lunch hour was to play a good close, competitive game.

Virtually anyone who entered the gym at lunchtime could be linked to one of these groups. For many of the students, playing in this forum was very public and quite risky. It was, however, an acknowledged proving ground for players who entered: Any aspiring basketball players had to prove themselves within a selected group before they would be accepted at a higher level.

Two underlying messages evolved from this venue: First, this was a male domain and the boys adopted several tactics to make sure people were where they wanted them to be. Second, if you wanted to be noticed by your peers, whether it be how you looked, what you were wearing or, more important, what you could do on the court, this was the place to do it.

A Similar Community Picture

Free-gym period at Shady Woods was similar to what occurred at the local recreation center. The after-school program started shortly after 2:30 p.m. By 2:45 the gym began to fill; it had two courts and six baskets. There were six basketballs available and the new arrivals raced to claim a ball. The following fieldnotes illustrate some of those similarities:

I am recognized by several of the youth who arrive at the center. "What you doin here?" they ask. The majority are dressed in shorts, T-shirt, and hightops. The center appears to be a place to "hang" as several of the kids congregate on the only row of the bleachers that is not retracted. On the second court the coordinator joins in on a game with seven older boys. He controls the game, coaching the players as well as playing himself. The seven girls in the gym have all retired to the bleachers. The kids on the bleachers converse among themselves about who did what or who should have done something else. The "big" game is very physical and played at a fast pace. The gallery increases in size and a new arrival calls, "Who's got next?" as he surveys who is in the gallery.

My note-taking catches the attention of a young woman who asks, "Are you NBA?" I smile at her and ask if she is going to play, to which she responds, "No they don't let girls play. . .they sexist!" whereupon she takes another swig of her Mountain Dew. The player who was marked by the coordinator voices his concerns that he has been pushed around by his opponent who is taller and considerably bigger than him. The young player's protests are met with. "You cryin all the time. . .you cry baby. . .Hey, we beat you by two. . .we beat you by two." The losers retire to the foyer to get a drink. (Fieldnotes)

The picture was similar across town at a center that also has Shady Woods students in attendance. Those attending were nearly all African American. The open-roll policy of the school meant that students could attend Shady Woods from throughout the city. The following fieldnote illustrates a similar ecology across town:

They organize pick-up games between themselves while the supervisor sits at the doorway and monitors behavior. She warns them, "The next cuss word means you are all out of here." Her warning is heeded. The center remains open until 10 p.m. Games dissolve into shooting contests or are transformed into another game almost without warning. Unlike their observer, the players seem to know how the system works. The young people advise me that they usually only play basketball and participate in the sport for most of the year. "Basketball is cool." (Fieldnotes)

Numerous visits to venues confirmed that the highly competitive nature of the games seemed to attract boys to organized and recreational sport more than girls. The intensity, physicality, and degree of "trashing" that occurred in many games we observed confirmed that the players often treated their games as though there was a lot at stake. This was evidenced at both recreation centers as well as at Shady Woods during lunch hour. Amanda, who has tried to understand why the boys take basketball so seriously, related her conversations with boys who play ball during lunchtime free gym:

They tell you, like, if they didn't do good or beat anybody, then they're not good enough for themselves so they get mad with themselves. And they sometimes try blaming it on other people when it's really their fault. I mean, they just have to accept that they can't and they have to do better.

The intensity that sometimes evolved from sporting encounters was not restricted to the boys. Jenna, who was one of the few Shady Woods girls to play ball during lunch hour, signaled a concern that,

There's a lot of people who take the games too serious and them's the ones you got to watch out for because they're the ones that will want to start a fight with you, want to pick a fight. . .but that's basically it.

It is possible that such demonstrations serve as "cool pose" (Patterson, 1997) resistance to or dissatisfaction with parts of the existing sport system in high school.

Preference for basketball was illustrated during daily open gym during the 5 months during lunch hour. It was the only sport ever played in the gym during the 5 months of observation at the school. This engagement with basketball, particularly by African American males, has been described as a "selective overemphasis of a dominant cultural norm" (Patterson, 1997, p. 188). Patterson drew on personal research of African American youth and discovered that "getting a young girl pregnant and being good at sports, or just being sports obsessed, are together the coolest things one can do and be" (p. 188). Basketball is cool at Shady Woods. However, it was available only to the few self-selected "cool-pose" participants who had the skills and confidence to control the situation. As Traci, a senior nonparticipant, said, "Nobody is going to say anything. They have their way and it's the same thing over and over again. Every day it's basketball, basketball."

The lunchtime power structure was dominated by skilled male African American basketball players. There was a clear expectation that everybody would conform to the norms of behavior established by this group. Most students were reluctant to challenge the power structure. We believe this situation is not unique to Shady Woods High School lunchtime or to Midwestern or urban schools.

During free-gym periods the basketball games were intense, competitive, and public. Anybody who wanted to play had to be good enough to get a place on a pick-up team and deal with attention from students who were watching. Many players were subjected to taunts and jeers from their peers. The power structure that evolved in the gym at Shady Woods privileged skilled, mostly African American male athletes while girls and less skilled boys were often excluded. While the issue of gender exclusion has been the focus of research in physical education environments (Bain, 1990; Carlson, 1995; Kollen, 1981), attention to informal sport settings has been meager. Even girls like Jenna, a competent basketball player, were reluctant to actively participate in free gym. If an environment is perceived as too public and marked by intense peer scrutiny, the perceived risks may outweigh the benefits of participation. By contrast, those who stayed were the skilled, boisterous, and aggressive. For them, success was virtually assured, as they had the necessary traits to survive. They were the advantaged and others had to adapt or retreat. The gym supported a differential survival whereby those who had power survived. It is a very negative example of a Darwinian existence.

In an educational setting, the characteristics of a Darwinian climate based on survival of the fittest are inappropriate. While we have conducted this study in light of the basic characteristics of Darwinism, we do not wish to enter into a theoretical discussion on Darwinian doctrine. We would, however, point to the need for an educationally sound alternative for how urban youth could evolve in a sport setting. For example, Depew (1996) suggested:

Perhaps it is not too much to say that what we need is an evolutionary theory worthy of our best social theory, not a theory trimmed to fit a rapidly receding, overly simplistic, evolutionary theory. For our part, we look forward to an ecologically grounded evolutionary theory whose point is the protection of individuals, communities, and their traditions in a natural world that is our true and only home. (p. 495)

However, it was evident that sport was very much a part of this youth culture and that it had an established power structure. Status could be gained through sport participation and approval from significant others. Observation of the free-gym situations indicated that the power players were the Bullies. They determined the space allocation, they attracted the biggest crowds, and they were the group that other participants aspired to become. As a Bully, you were expected to be an extremely competent player and learn the local culture, its conventions and its characteristics, in essence almost like an apprenticeship.

Learning by Apprenticeship

In a study that explored elements of successful programs, Resnick (1987) found that the notion of apprenticeship as a means of learning "allows skill to build up bit by bit, yet permits participation even for the relatively unskilled, often as a result of social sharing of tasks" (p. 18). Lave and Wenger (1991) suggest that "learners inevitably participate in communities of practitioners and that the mastery of knowledge and skill requires newcomers to move toward full participation in the socio-cultural practices of the community" (p. 29).

Before full participation can occur, apprenticeship is often served, requiring the novice learner to indulge in a limited form of legitimate peripheral participation. As he or she serves his/her apprenticeship, "a person's intentions to learn are engaged and the meaning is configured through the process of becoming a full participant in the socio-cultural practice. This social process includes, indeed it subsumes, the learning of knowledgeable skills" (Lave & Wenger, 1991, p. 29). This model of learning holds true for the ecology of the free-gym environment. Many students had to "do their time" on the periphery of the court before they could gain access to games of increasing complexity and status. But it was not only the game and its associated skills that had to be learned. Ambitious and hopeful players had to learn and accept the social conventions associated with each game of basketball. In many cases, to learn the game's social conventions was more important than the performance itself. Such a process promotes one's survival as a future participant. These attitudes and behaviors are a reflection of street literacy and leave no room for less able and less aggressive students to participate in a positive physical activity experience.

Street Literacy and the Ways of Youth

The production of informal knowledge from practices that occur within contexts such as free gym has been referred to as street literacy. Cahill (2000) suggests that "street literacy is an interpretive framework that privileges experienced informal local knowledges that are grounded in personal experiences and passed down in the form of rules, boundaries set by parents, neighborhood folklore, and kids' collective wisdom" (p. 252). She argues that urban learning sites such as pick-up basketball venues hold considerable significance for young people who attend such contexts. It is within such specific contexts that urban youth can construct knowledge based on personal and social experiences.

This perspective is supported by Resnick (1987), who argues that "work, personal life, and recreation take place within social systems, and each person's ability to function successfully depends on what others do and how several individuals' mental and physical performances mesh" (p. 13). Moreover, each individual must compete not only with the performances of others but also "with larger institutional, social, and economic contexts" (Cahill, 2000, p. 271). Urban youth are therefore the holders of multiple worlds that intermesh and compete, thereby influencing how each young person expresses himself/herself.

While free gym can he a forum in which many urban youth have the opportunity for self-expression, the dynamics at Shady Woods would indicate that for most students the opportunities are nonexistent. The social machinery evident at Shady Woods reflects Larry Locke's observation that "It is easy to believe, however, that student influence is both great and sometimes destructive to educational purposes" (1974, p. 15). According to Katz (1995), it would appear that the ecology of free gym is marked by power, space, and terror.

Katz (1995) reports that while many outdoor environments suffer from "benign neglect," young people are increasingly seeking the safety of indoor spaces. Moreover, what play that remains in outdoor leisure environments is repeatedly taken over by boys, with girls often being restricted from such unsupervised spaces. The school thus becomes one of the few preserves for youth seeking active leisure. Sadly, our research would suggest that while the venue may have changed, the educational setting itself has fallen victim to benign neglect. Our prolonged presence at Shady Woods suggested that the Darwinian existence of free gym received the tacit approval of staff. Those who supervised the lunch-hour sessions indicated that it served as a cathartic outlet for students and that it was not advisable for them as supervisors to rock the boat.

The question one must ask is, "What are we, as physical educators, going to do about it?" Clearly the nature of free gym has implications for physical education. It is perhaps indicative of the challenge that physical education teachers must face between school culture and that of the wider society. At a micropolitical level, any attempt by teachers to change classroom culture will first require attention to and knowledge of those wider cultural influences (such as street literacy) as significant determinants of what occurs in the gym during physical education.

The changing contexts of the youth sport culture have implications for how sport can and should be delivered or, more important, how it has been delivered. If youth are to be given options for participation, those options must be viewed as viable, attractive, appropriate, and enjoyable. Tinning and Fitzclarence (1992) have argued that for postmodern youth culture, "the disjunction between physical education and the place of physical activity in their out-of-school lives is contradictory" (p. 301). This perspective is endorsed by Csikszentmihalyi (1993), who argued that "we bring up children to take their places in a culture that, in reality, no longer exists. The basic skills they learn have little to do with survival in the future" (p. 276). While free gym occurred during lunch hour, the realities of such a setting appeared to have an impact on the ecology of physical education classes (see O'Sullivan, Tannehill, Knop, Pope, & Henninger, 1999; Pope & O'Sullivan, 1998).

Students who were marginalized by the free-gym ecology reported for physical education classes at the same venue and often alongside the same students who either controlled or benefited from how life was ordered during lunch hour. The implication suggests that many students would be required to learn and embrace a different ecology for physical education, while at the same time living with the powerful and pervasive student-centered alternative during noninstructional school time.

If we as educators remove ourselves from ecologies such as free gym, only a few young people will be allowed to survive and learn what appear to be desirable skills in a culture of convenience for a small but powerful group. Teachers and students may need to find some middle ground between these ecologies. Brown and Theobald (1998) argued, "the aim of these [extracurricular] activities, therefore, should be to meet all students' needs, to enrich their learning experiences, rather than to create a small cadre of exceptionally skilled youth" (p. 133). As teachers, we have a moral responsibility to arrest socially chronic situations such as free gym. While such a context is "free" of teacher authority, it cannot really be termed free unless all restraint, including that which is placed by dominant peers upon their less dominant peers, is removed.

Today's youth value social affiliation, and a context like free gym has the potential to promote something that is inextricably attractive to them. To realize that potential, a compatible pedagogy must be employed. This would not mean building from scratch. Social development (Hellison, 1995; Hellison, Cutforth, Kallusky, et al., 2000) and sport education (Siedentop, 1994) models have already emerged as viable options to promote sport as an inclusive physical education and sport environment.

Traditional ethnography has often explored the difference between ideal and real culture (Angrosino & Mays de Perez, 2000). The ecology at Shady Woods illustrates the "real" which, as educators, we could compare to our perception of the ideal sporting environment that is educational and marked by enjoyment, learning, and access for all who seek it. The ideal/real contrast is something that many teachers and teacher educators are faced with. It required us to position ourselves as white middle class educators anchored in an environment dominated by urban male African American youth. We cannot assume an ideal, nor can we assume a universal meaning because meaning shifts from one ecology to another.

Although this study reflects the ecology of one extracurricular context, some of the issues raised are similar to those experienced elsewhere. In England, Penney and Harris (1997) argue, there is a need to reevaluate the provision and content of extracurricular activity. Similarly, James (1999) has highlighted the dominance of skilled Australian boys in casual basketball settings who exclude and dissuade others, particularly girls, from participating in the activity. While we may not have the immediate answers, it is important, as Willard-Holt (2000) has contended, to pay credence to the lived experiences of young urbanites to thereby provide a meaningful education for them.

Conclusion

To ameliorate the delivery of sport in urban high schools, teachers need to acknowledge the impact of street literacy. In so doing, they may have some hope of moving from the real toward the ideal. Ethnographies can reveal a great deal about the ecology of the gym, but do we merely report ethnographies that give readers a perspective of how it is? Or do we take it one step further and seek to create social or educational change because the universal rights of some participants (girls and less skilled boys) are marginalized?

Sport is arguably a pervasive characteristic of everyday urban life. More important, it forms a significant part of school life. The question we must ask is, If sport is situated within a school context, is it educational? Gerdy (2000) has warned us that "sport is becoming so disconnected from educational values and foundations that it is in grave danger of becoming educationally irrelevant" (p. xi). Although free gym thrives in many urban schools, it is rarely part of the wider educational debate. It is evident from this study that many of the potential qualities that sport could offer as an educational medium have either been diminished or lost.

David Hamburg of The Carnegie Council on Adolescent Development stresses that for adolescents to make the transition to adulthood, they will require "an array of stimulating, constructive opportunities throughout their waking hours—both in school and beyond" (1997, p. 11). Accordingly, there needs to be a place where interesting programs are available to those who seek health and enjoyment through sport. The challenge for teachers is to ascertain whether the suitability of participant experiences in certain sport forms is meeting the developmental and social needs of many young people. Any effort to enhance the wider sport culture should include paying credence to the experiences young people have with sport.

References

Allison, P.C. (1988). Strategies for observing during field experiences. *Journal of Physical Education, Recreation and Dance, 59*(2), 28-31.

Angrosino, MV., & Mays de Perez, K.A. (2000). Rethinking observation. In N.K. Denzin & Y.S. Lincoln. (Eds), *Handbook of qualitative research* (2nd ed., pp. 673-702). Thousand Oaks, CA: Sage.

Bain. L. (1990). A critical analysis of the hidden curriculum. In D. Kirk & R. Tinning (Eds.), *Physical education, curriculum and culture* (pp. 23-42). London: Falmer Press.

Berk, L.E. (1992). The extracurriculum. In P.W. Jackson (Ed.), *Handbook of research on curriculum* (pp. 1002-1043). New York: Macmillan.

Brettschneider, W.D. (1990). Adolescents, leisure, sport and Lifestyle. In T. Williams, L. Almond, & A. Sparkes (Eds.), *Sport and physical activity: Moving towards excellence* (pp. 536-550). London: E & FN Spon.

Bronfenbrenner, U. (1979). *The ecology of human development: Experiments by nature and design.* Cambridge, MA: Harvard University Press.

Bronfenbrenner, U. (1989). Ecological systems theory. *Annals of Child Development, 6,* 187-249.

Brown, B.B., & Theobold, W. (1998). Learning contexts beyond the classroom: Extracurricular activities, community organizations, and peer groups. In K. Borman & B. Schneider (Eds.), *The adolescent years: Social influences and educational challenges* (pp. 109-141). Chicago: University of Chicago Press.

Cahill, C. (2000). Street literacy: Urban teenagers' strategies for negotiating their neighbourhood. *Journal of Youth Studies, 3,* 251-277.

Carlson, T. (1995). We hate gym: Student alienation from physical education. *Journal of Teaching in Physical Education, 14,* 467-477.

Csikszentmihalyi, M. (1993). *The evolving self.* New York: Harper Perennial.

Cusick, P.A. (1973). *Inside high school: The student's world.* New York: Holt, Rinehart &Winston.

Denzin, N. (Ed.) (1989). *The research act: A theoretical introduction to sociological methods.* Englewood Cliffs, NJ: Prentice Hall.

Depew, D. (1996). *Darwinism evolving: Systems dynamics and the genealogy of natural selection.* Cambridge, MA: MIT Press.

Eder. D., & Parker, S. (1987). The cultural production and reproduction of gender: The effect of extracurricular activities on peer-group culture. *Sociology of Education, 60,* 200-213.

Eitzen, D.S. (1995). Classism in sport: The powerless bear the burden. *Journal of Sport and Social Issues*, 20, 95-105.

Evans, J., & Roberts, G.C. (1987). Physical competence and the development of children's peer relations. *Quest*, 39, 23-35.

Ewing, M.E., Seefeldt, V.D., & Brown, T.P. (1996). *Role of organized sport in the education and health of American children and youth*. East Lansing, MI: Institute for the Study of Youth Sports, Michigan State University.

Garton, A.F., & Pratt, C. (1987). Participation and interest in leisure activities by adolescent school-children. *Journal of Adolescence*, 10, 341-351.

Garton, A.F., & Pratt, C. (1994). Leisure activities of adolescent school students, predictors of participation and interest. *Journal of Adolescence*, 14, 305-321.

Gerdy, J.R. (Ed.) (2000). *Sports in school: The future of an institution*. New Your: Teachers College Press.

Griffin, P.S. (1984). Girls' participation patterns in a middle school team sports unit. *Journal of Teaching in Physical Education*, 4, 30-38.

Griffin, P.S. (1985). Boys' participation patterns in a middle school team sports unit. *Journal of Teaching in Physical Education*, 4, 100-110.

Grupe, O., & Kruger, M. (1994). Sport pedagogy: The anthropological approach. *Sport Science Review*, 3(1) 18-27.

Hamburg, D.A. (1997). Meeting the essential requirements for healthy adolescent development in a transforming world. In D.A. Takanishi & R. Hamburg (Eds.), *Preparing adolescents for the twenty-first century* (pp. 1-12). Cambridge: Cambridge University Press.

Hansen, J.F. (1979). *Sociocultural perspectives on human learning: An introduction to educational learning*. Englewood Cliffs, NJ: Prentice Hall.

Hellison, D. (1995). *Teaching responsibility through physical activity*. Champaign, IL: Human Kinetics.

Hellison, D., Cutforth, N., Kallusky, J., Martinek, T., Parker, M., & Stiehl, J. (Eds.) (2000). *Youth development and physical activity*. Champaign, IL: Human Kinetics.

James, K. (1999). "I feel really embarrassed in front of the guys!" Adolescent girls and informal school basketball. *The ACHPER Healthy Lifestyles Journal*, 46(4), 11-16.

Katz, C. (1995). *Power, space and terror: Social reproduction and the public environment*. Presented at the Landscape Architecture, Social Ideology and the Politics of Place conference. Harvard University, Cambridge, MA. Unpublished paper.

Kollen, R (1981). *The experience of movement in physical education: A phenomenology*. Unpublished doctoral dissertation, University of Michigan.

Kozol, J. (1991). *Savage inequalities*. New York: Crown Publishers.

Lave, J., & Wenger E. (1991). *Situated learning: Legitimate peripheral participation*. Cambridge: Cambridge University Press.

Lincoln, Y., & Guba, E. (1985). *Naturalistic inquiry*. Beverly Hills, CA: Sage.

Locke, L.F. (1974). *The ecology of the gymnasium: What the tourists never see*. SAPECW Workshop, Gatlinburg, TN. (Eric Document No. ED 104 823)

O'Sullivan, M., Tannehill, D., Knop, N., Pope, C., & Henninger, M. (1999). A school-university collaborative journey toward relevance and meaning in an urban high school physical education program. *Quest*, 51, 225-243.

Patterson, 0. (1997). *The ordeal of integration*. Washington, DC: Civitas.

Patton, M.Q. (2001). *Qualitative research and evaluation methods* (3rd ed.). Thousand Oaks, CA: Sage.

Penney, D., & Harris J. (1997). Extra-curricular physical education: More of the same for the more able? *Sport Education and Society,* 2, 41-54.

Pope, C., & O'Sullivan, M. (1998). Culture, pedagogy and teacher change in an urban high school: How do you want your eggs done? *Sport Education and Society,* 3. 201-226.

Resnick, L.B. (1987). Learning in and out of school. *Educational Researcher,* 16(9), 13-20.

Schempp. P. (1998). The dynamics of human diversity in sport pedagogy scholarship. *Sociology of Sport On-line* [On-line. Retrieved July 30, 2001]. Available: http://physed.otago.ac.nz/sosoUvlil/vlila8.htm

Siedentop, D. (1994). *Quality PE through positive sport experiences: Sport education.* Champaign, IL: Human Kinetics.

Siedentop, D. (1996). Valuing the physically active life: Contemporary and future directions. *Quest,* 48, 266-274.

Spradley, J.P. (1980). *Participant observation.* New York: Holt, Rinehart & Winston.

Tinning, R., & Fitzclarence L. (1992). Postmodern youth culture and the crisis in Australian secondary school physical education. *Quest,* 44, 287-303.

Werner, 0., & Schoepfle, G.M. (1987). *Systematic fieldwork: Vol. 1. Foundations of ethnography and interviewing.* Newbury Park, CA: Sage.

Willard-Holt, C. (2000, March). Preparing teachers for urban settings: Changing teacher education by changing ourselves. *The Qualitative Report,* 4(3/4), Article 05. Retrieved March 31. 2001, front http://www.nova.edu/ssss/QR/QR4-1/willard.html

Appendix C

Physical Activity Experiences of Women Aging With Disabilities

Donna L. Goodwin and Scott G. Compton
University of Saskatchewan

This hermeneutic phenomenological study sought to understand the experiences of physical activity and aging with a disability. Six women with physical disabilities, including cerebral palsy (n = 2), acquired brain injury (n = 1), and spinal cord injury (n = 3), and between the ages of 22-37 years (mean age = 28 years) participated in the study. Their experiences were captured by way of semi-structured interviews. Each participant completed two interviews that were audiotaped and transcribed verbatim. The thematic analysis revealed three themes: experiencing something normal, loss of physical freedom, and maintaining function through physical activity. Implications of the findings were discussed within the context of health promotion and Verbrugge and Jette's (1994) socio-medical model of disablement.

Aging with a disability and the efficacious effects of physical activity on health and function has received relatively little attention from adapted physical activity researchers. And yet, the number of persons with early onset disability who are reaching middle and old age is increasing (Vandenakker & Glass, 2001). The advent of antibiotics, improved acute and rehabilitative care, and the prevention and treatment of secondary disabilities has contributed to more persons surviving injury and enjoying longer life expectancies (Kemp & Miosqueda, 1997). For example, in contrast to the 1950s when the life expectancy of a person with a spinal cord injury was about 5 years post injury, today's life expectancy is approximately 85% that of a person without a major disability (Kemp & Miosqueda, 1997; Lammertse & Yarkony, 1991). Nevertheless, health concerns increase as more individuals continue into their second, third, and fourth decade beyond injury (Charlifue, Weitzenkamp, & Whiteneck, 1999).

The Aging Process

The aging process itself may be accelerated or intensified by the cumulative stress of repetitive activities and altered body mechanics that can occur with disability. Whereas persons with disabilities have been led to believe that their impairments were stable, unexpected changes in functional ability are appearing 20 to 25 years after the onset of

Entire text of appendix C is reprinted, by permission, from D.L. Goodwin and S.G. Compton, 2004, "Physical activity experiences of women aging with disabilities," *Adapted Physical Activity Quarterly* 21: 122-138.

the injury, often between the ages of 30 to 50 (Vandenakker & Glass, 2001). This suggests that persons with physical disabilities commonly encounter problems associated with aging 10-20 years earlier than persons without disabilities (Whiteneck et al., I 992).

Upper extremity pain and degenerative changes of the shoulder are documented in persons with disabilities (Lal, 1998). In a study of 53 people with spinal cord injuries of at least 15 years duration, Lal (1998) found that 78% of the participants demonstrated radiological degenerative evidence of changes to their shoulders. Those who used wheelchairs and the women of the group were most susceptible to these changes.

Pain may accompany changes in function. Pentland, McColl, and Rosenthal (1995) found that with increasing age, persons with spinal cord injuries decreased activity levels due to pain. Self-efficacy for activity may well be impeded if participation in physical activity results in an increase in pain, fatigue, or stiffness (Dunlop, Hughes, Edelman, Singer, & Chang, 1998; Lal, 1998). The physical inactivity that accompanies joint pain is a further predictor of a decline in function, the ability to complete activities of daily living, and therefore personal independence (Dunlop et al.,1998; Gerhart, Bergstrom, Charlifue, Menter, & Whiteneck, 1993).

Physical Activity as an Intervention

With age, health concerns such as hypertension, arthritis, osteoporosis, heart disease, and diabetes increase (Kemp & Miosqucda, 1997). Persons with disabilities may be at even higher risk of these health concerns due to sedentary lifestyles, lack of support systems, underemployment, and unique manifestations of symptoms that may not have received immediate diagnostic attention (Cooper et al., 1999).

The importance of physical activity in maintaining and enhancing physical function as we age has been well documented (e.g., Carlson et al., 1999: Ettinger, 1998; Hootman, Sniezek, & Helmick, 2002; Kavanagh & Shephard, 1990). Its role in the prevention of disabling secondary conditions such as stroke, coronary artery disease, and diabetes in persons with primary disabilities has also become apparent (Rejeski & Focht, 2002). Physical activity can increase strength, which serves to maintain and increase function, which in turn can affect normal physiological functions and thereby indirectly limit the risk of potentially disabling diseases in aging adults with disabilities (Carlson et al., 1999).

Much of the physical activity intervention research aimed at preventing secondary disabilities has occurred in institutional settings with participants who, because of their age, have difficulty completing the exercise protocols, fail to complete the intervention due to illness, or drop out (Carlson et al., 1999). The participants of these studies often have age related primary disabilities rather than those acquired through injury or impairment earlier in life. Nonetheless, the mediating effect of physical activity on disability decline is apparent.

Hirvensalo, Rantanen, and Heikkinen (2000) followed 1,109 men and women between the ages of 65 and 84 years who lived independently in the community. Over an eight year period, 32 men and 125 women became dependent, meaning they were required to move to a nursing home, hospital, or other assisted-living facility. The participants were grouped according to their mobility and activity levels: intact mobility and active, intact mobility and sedentary, impaired mobility and active, and impaired mobility and sedentary. Intact mobility meant the person could walk two kilometers and climb one flight of stairs without difficulty. Difficulty completing one or both of the activities resulted in an impaired mobility designation. The impaired active group and the impaired sedentary group were, respectively, three and five times more likely to become dependent than the mobile active group over the same period of time. The authors concluded that physical activity appeared to protect people with mobility impairments from further disability over those with mobility impairments but who were sedentary.

In a sample of noninstitutionalized individuals, Miller, Rejeski, Reboussin, Ten-Have, and Ettinger (2000) surveyed 5,151 men and women, 70 years and older in 1984, 1986, 1988, and 1990. They concluded that physical activity could slow the progression of functional impairments commonly associated with older age and the transition to disability. Disability was defined as the inability to complete activities of daily living to the point that assistance was needed to get in and out of bed, shop, or complete light housework. The results reinforced the position that those who participated in physical activity, defined as walking a mile at least once per week, had a greater probability of improving or not progressing in age-related functional limitations and disability over their sedentary counterparts. The analysis further revealed that the relationship stood irrespective of whether the participants had preexisting severe functional limitations or disabilities.

The Meaning of Aging

In addition to our understanding of the biological consequences, consideration must also be given to the psychosocial aspects of aging with a disability. Metaphors pertaining to the body, time, and others often capture our feelings, fears, and adjustment to aging (Adams-Price, Henley, & Hale, 1998). Metaphors that describe physical changes are abundant. We make reference to the gray-haired crowd, admonish those who are as wrinkled as a prune, and refer to those who use canes or walkers as old geezers. Aging has been referred to as a race against time and many of us attempt to mask the ravages of time. Older adults may also be heard to say "my time has come," while others may comment, "it was her time" (Adams-Price et al., 1998). And yet, denial of aging has been viewed as being instrumental in the promotion of successful aging. The adage, "You are only as old as you feel" speaks to this sentiment (Montepare & Lachman, 1989).

Defining oneself as younger than actual age has been shown to be positively related to psychological adjustment and health (Montepare & Lachman, 1989). Researchers have reported that the vast majority of older people even those over 80 years of age, feel considerably younger than their years (Adams-Price et al., 1998). There is evidence that only those in poor health actually consider themselves old, but that this may be due in part to being treated poorly by caregivers or medical professionals (Keller, Leventhal, & Larson, 1989).

For persons with disabilities, there have been inconsistencies in the reported relationships among increasing age, time since injury, and subjective well-being and life satisfaction. In a summary article, Eisenberg and Saltz (1991) concluded that the quality of life enjoyed by persons with spinal cord injuries, young and old, could be relatively good. In fact, for veterans with spinal cord injuries, quality of life was reported to actually be better than similarly aged males without disabilities. In contrast, a study by Wheeler, Malone, VanVlack, Nelson, and Steadward (1996) on the retirement of elite athletes with disabilities highlighted the worry associated with aging. Not knowing what to expect, losing function and moving to an electric wheelchair, and becoming dependent upon others were sentiments expressed by the participants.

Conceptual Framework

The disablement process is a term that has been used to describe the biomedical, social, and political schemas that delineate the transition from health to disability (Lawrence & Jette, 1996). Early disablement models upheld a medical and individualistic approach to disability, claiming disability to be an outcome of bodily pathology (e.g., Nagi, 1965). Hughes and Paterson (1997) summarized the evolution that has occurred within disablement models: "The medical view that social restrictions for disabled people were a consequence of physical dysfunctions was overturned by a radical move which argued that people with impairments were disabled by a social system which erected barriers to their participation" (p. 328). Current models of disablement endeavor to integrate the impact of personal, social, and environmental factors on physical function (e.g., Hughes & Paterson, 1997; Peters, 1996; Verbrugge & Jette, 1994). This shift in philosophy is also reflected in the International Classification of Functioning developed by the World Health Organization in 2002.

Verbrugge and Jette (1994) presented a sociomedical model of disablement. They describe how chronic and acute conditions can affect the functioning of specific body systems, physical and mental actions, and activities of daily living. The authors assert that the disablement process begins with a medically labeled pathology based on the presence of a biochemical or physiological abnormality, such as a disease mechanism, an injury, or a congenital or developmental condition. Aging presents various pathologies to systems of the body and therefore can begin the process of disablement (Boult, Kane, Louis, Boult, & McCaffrey, 1994).

The labeled pathology can result in impairments or anatomical and structural abnormalities to specific body systems. Impairments can lead the way to functional limitations or restrictions in basic physical and mental actions used in daily life. Functional limitations can ultimately lead to difficulty completing activities in different domains of life and hence disability. Functional limitations have been described as what a person can do and disability is what the person does do, thereby suggesting that functional limitations and disability are two aspects of the same behavior (Verbrugge & Jette, 1994). Verbrugge and Jette's (1994) model of disablement suggests that personal and environmental factors can influence, avoid, or hasten the gap between personal capacity and environmental demands. It has been suggested that physical activity can maintain a restoring function and hence mediate the final outcome of the disablement process (Hopman-Rock, Kraaimaat, & Bijlsam, 1996; Pentland et al., 1995).

The dynamic interactions of interpersonal, social, and environmental factors that can influence the impact of impairment on function and ultimately disability provides a useful heuristic framework from which to understand and interpret the experiences of physical activity and aging with a disability. As a result, Verbrugge and Jette's (1994) model of disablement provided the conceptual framework from which the findings of this study were interpreted.

Purpose

The purpose of the study was to gain an understanding of the aging experiences of young women with physical disabilities. More specifically, the objectives were to (a) understand the bodily experiences of aging in young women with physical disabilities, (b) uncover the meaning ascribed to those experiences within the context of physical activity, and (c) give voice to the health concerns of women with physical disabilities.

Method

Hermeneutic phenomenology was the research approach utilized. It is the study of the interpretation of texts for the purpose of obtaining a common understanding of the meaning assigned to everyday experiences (van Manen, 1997). It provided the sensitive and sophisticated perspective needed to examine the embodied experiences of women with disabilities (Paterson & Hughes, 1999). Phenomenology offers a descriptive, reflective, interpretive, and engaged mode of inquiry that seeks to understand and describe the essence of experiences and enables underlying structures (themes) and commonalties in meanings to be understood (Cresswell, 1998; Moustakas, 1994). The approach was hermeneutic in nature because gaining an understanding of the commonality of meaning behind the experiences requires interpretation on the part of the researcher and reader (Allen & Jensen, 1990).

The data presented were from a larger study (Goodwin, Kuttai, & Harris, 2001). The purpose of the original study was to describe the relationship between women's experiences of disability and physical activity. The participants were asked to share their stories about their bodies and how they described themselves, what they did during the week that they would term physical activity and the role it played in their lives, and the meaning that physical activity held for them. Although the topic of aging with a disability was not the subject of the original study nor was it probed for during the interviews, 6 of the 10 women spontaneously and openly discussed their concerns, fears, and apprehensions about aging. This was of considerable surprise to the investigators as the mean age of the participants was 30 years (range 22-49 years). For this reason, it was felt that the experiences of these six women needed to be brought to the attention of adapted physical activity researchers and those interested in women's health. Physical activity and aging with a disability and has received relatively little research attention.

Participants

A maximum variation purposeful sampling design was utilized, meaning that within the participant group of interest, representation was sought from a broad range of its members (Patton, 1997). This sampling strategy was particularly useful for capturing central themes or principal outcomes that cut across the experiences of women with different physical disabilities (Dreher, 1994). The common patterns that emerged are thereby central or core to their experiences as women with disabilities, irrespective of their disability classification.

Six caucasian women with physical disabilities with a mean age of 28 years (range 22-37 years) are represented in this study. Congenital and acquired physical disabilities were represented within the participant group and included cerebral palsy (n = 2), spinal cord injury (n = 3), and acquired brain injury with muscular weakness (n = 1; see Table 1). The women were recruited with the support of a provincial wheelchair sport organization. The women were all employed and described themselves as physically active. Some women participated in activities such as swimming, aerobics, biking, walking or wheeling, and weight training, while others considered cooking meals, doing housework, getting ready to go out (showering, dressing, preparing for an evening out at the movies), and purchasing groceries as physical activity given the nature of their disabilities.

Data Collection

The participants' experiences were captured by way of audio taped one-on-one, semi-structured interviews and field notes (Kvale, 1996). Each participant completed two interviews within a two-week period, each lasting between 60 and 90 minutes. After the first interview, the participants were asked to provide a photograph from their personal albums that further illustrated their experiences. The significance of the images was discussed during the second interview. The second interview also afforded the opportunity to explore questions not covered in the first interview and probe further into the stories shared by the participants. The interview guide included questions such as tell me about yourself, what does physical wellness mean to you, what value/role do you place on physical wellness in your life, and how does participation in physical activity impact how you view yourself?

Table 1 Description of Participants

NAME	AGE[1]	DISABILITY[2]	MOBILITY[3]	OCCUPATION	PHYSICAL ACTIVITY INTERESTS
Dawn	31	SCI	Manual WC	Educational Counselor	Air rifle shooting, swimming, camping, wheeling
Amanda	37	ABI	Ambulatory	Seniors' Aide Worker	Biking, walking, swimming, stretching, weights at home
Leanne	22	CP	Power WC	Educational Counselor	Activities of daily living
Jennifer	27	SCI Incomplete Quadriplegia	Ambulatory	Teacher	Biking, swimming, walking, weight training
Cathy	22	SCI Paraplegia	Manual WC	University Student	Wheeling, activities of daily living
Sarah	29	CP	Ambulatory	Social Services Employee	Swimming, cycling, kayaking, skating, horseback riding

[1]Age in years

[2]ABI = Acquired Brain Injury, CP = Cerebral Palsy, SCI = Spinal Cord Injury

[3]WC = Wheelchair

Field notes were recorded after the interviews, not only to record events within the interview environment as would be seen through the lens of a camera, but to also capture the reflections, hunches, and emotions of the interviewers (Jackson, 1990). The field notes were entered as research data and facilitated a conceptual return to the interview setting during data analysis (Creswell, 1998). Data analysis began after both interviews were completed. It was during the process of taking field notes that the topic of aging was noted.

Data Analysis

To identify common threads that extended throughout the data, an inductive analytic thematic analysis was conducted (van Manen, 1997). To isolate the emerging thematic statements, a line-by-line analysis was conducted which entailed reading the transcripts and field notes numerous times. Particularly revealing phrases were highlighted and coded with meaningful labels (e.g., pain, loss of function, concern for future). The data analysis continued by constantly comparing phrases to determine whether they should be classified separately or whether they belonged to an existing code (Wolcott, 1994, 2001). Those that were conceptually similar were gathered together into thematic statements and supported by quotes from the participants.

Trustworthiness

Rigor was incorporated into the design of the original study by way of verification (the extent to which the essences of the experiences were captured), validation (the procedures undertaken and the information presented were compatible with the research question), and confirmability (or the degree to which there is little reason to doubt the truth of the findings; Denzin, 1994; Meadows & Morse, 2001). The interview guide was developed with the support of a woman with a physical disability, thereby enhancing the verification to the findings (Meadows & Morse, 2001). Her insider perspective brought credibility to the nature of the questions asked and their relevancy to the research question posed in the original study. (Gill, 1997; Peters, 1996).

Theoretical triangulation also contributed to the verification or credibility of the study. The research question was situated within current literature pertaining to the experience of disability, aging with a disability, and the process of disablement (Peters, 1995, 1996; Rantanen et al., 1999; Rejeski & Focht, 2002; van Manen, 1997; Verbrugge & Jette, 1994).

To enhance the validity or dependability of the findings, two data-coders (authors) were employed. The codes and labels were verified between the first and second author and the themes that emerged were collaboratively determined (Meadows & Morse, 2001).

To reduce researcher bias and enhance the confirmability of the findings, investigator triangulation was undertaken (Cresswell, 1998). The research team (interviewer, principal investigator, and contributing author) all possessed an adapted physical activity background as well as selected expertise in qualitative inquiry, interview technique, and an interest in women's health.

The relevance and applicability of the results of this study to other women with physical disabilities rests with naturalistic generalization or the degree to which two contexts are similar (Denzin, 1994; Lincoln & Guba, 1985). The transferability or fittingness of the findings beyond the experiences described by the women of this study was enhanced by using participant quotes to illustrate the themes in addition to information on their age, race, disabilities, life occupations, and physical activity interests of the women. This information provides a solid foundation for recognizable congruence of women's experiences within other contexts (Schofield, 1990).

Results

Three themes emerged from the analysis of the data: experiencing something normal, loss of physical freedom, and maintaining function through physical activity. Sarah was the most articulate participant on the issue of aging with a disability and a number of her stories have been used to illustrate the themes. This should not be interpreted to mean that the themes presented are not representative of the other women's experiences. The themes arose from common experiences expressed by all of the participants. Before presenting the themes in more detail, Sarah provides a profound context from which to hear the stories of all the women.

> I am glad that I have a disability and that I am able to age. . .I mean there was times when I looked and I could not see anyone disabled who was healthy and older. And if I have to be one of the first that's publicly out there, then let it be. Sarah, age 29

Experiencing Something Normal

Sarah, who has cerebral palsy, spoke very eloquently about perceptions of perfectionism and of the good parts and bad parts of her body. She placed differential value on her body parts depending upon whether or not they were affected by cerebral palsy. With the exterior changes that occur to the body with age, however, Sarah was able to assimilate the good and bad into an integrated and self-validated whole and share in an experience common to all women.

> It's so much different than it was before, because I saw my body as two almost separate spheres with a kind of interconnect. When I was younger, I saw myself as young and beautiful and vivacious and everything was in tone, and then I had my legs to contend with. So I saw them as perfection and then imperfection. And then now I see myself as sort of all integrated, which because I am aging with a disability, and I actually see parts of myself that aren't disabled aging and I say, "Well, that's not perfect either," so it's sort of an amalgamation of the whole body. . . If you're not wrestling with your body in the sense of denying it or being embarrassed by it, of hiding from it, spiritually you will grow, emotionally you will grow because it all comes together. I think I am more comfortable because I had to come to terms with imperfection a lot sooner. And it comes back to how I see myself now, aging, like how the separate spheres are integrated and I see my whole body as aging, but I do see that as a normal healthy process. When I look at my face and my upper body, I see it aging just as well as my legs are, and I laugh because I think for once in my life I am experiencing something normal. Sarah, age 29

The women expressed a sense of pride in their physical accomplishments. They beat the odds in some respects, surpassing the expectations of physicians and parents. Nonetheless, the natural progression of completing school, career, love, marriage, and family is not necessarily applied to women with disabilities (Gill, 1997). The wisdom of beginning a family was questioned by one of the participant's mothers. She advised her daughter who had a spinal cord injury to not become pregnant out of fear for her physical health. "My Mom really didn't want me to have a baby" (Dawn, age 31). Taking her mother's concerns under advisement, Dawn nevertheless conceived a baby and a healthy boy was born to the couple through natural childbirth. Dawn went on to say that "I know now, more than I ever have before, that my body is capable of things that were never expected of me, that I didn't expect of myself. I learned that when I had the baby."

The psychological impact of age on the sorting out or integrating of feelings, perceptions of self, disability, and being a woman was clearly cherished. Dawn conveyed that her life experiences have been framed by the injury she sustained as a young girl and that life was a celebration because of the self-development that has resulted. She rejected the notion that she should down play her disability and "normalize" her experiences. A sense of self-assurance and calm about her future as a woman with a disability was evident. She gladly anticipated the unfolding of her life story in the decades to come.

I am not going to talk about or celebrate how good my life is in spite of [my injury]. I'm going to celebrate how good my life is because of my injury. I was really looking forward to turning 30, because I thought that things would sort of come together for me then, and I think they have. And I can't wait to know what I'm going to be thinking when I'm 40 and 50. I think I'm only going to get more introspective. Dawn, age 31

The women of this study were forced to deal with imperfections of the body that many women associate with aging at a much younger age. The maturity with which they approached aging was evident in their anticipation of experiencing something that all women experience.

Loss of Physical Freedom

The women were confident in their physical strength given the obstacles that they had overcome in meeting the challenges of maintaining an engaged and active lifestyle. The objectification brought to the body through medical diagnosis, surgery, rehabilitation, and ongoing monitoring of growth and development created a backdrop of experience that appears to have motivated the women to achieve in spite of current wisdom and to create and maintain a sense of self through their acquired independence. Goal setting for these women was done on their own terms and with tremendous commitment to their independence. Sarah shares the passion she feels toward the importance of physical well-being in her life and her battle to maintain a focus on health and physical well-being.

Sometimes I am afraid of my physical future. I feel like everything I've gained physically in my life has been because I have earned it. I just feel very passionately when I talk about physical well-being, because I just feel like it was almost robbed from me. I really feel like it was because I was so medicalized when I was a child, and I feel like my life wasn't seen outside of that little bubble. . .I hate x-rays, like I feel sick after x-rays, because I feel it's (osteoarthritis) taking away my physical freedom, because I have to balance my physical freedom against the potential for injury when I'm doing something new or something that I've always done. I am trying to prevent further injury, because I know that with further injuries that they are most likely going to be permanent. Sarah, age 29

The importance of independence to the women was clearly articulated. Their sense of personal happiness, freedom, and autonomy were all linked to using their bodies in ways that contributed to their sense of self. Learning, mastering, and maintaining independence in their lives was of primary concern. Dawn (age 31) acknowledged that in time she may need additional support with her mobility needs but wanted to enjoy the use of her body for as long as possible in that regard. "I really don't want to have an electric or power chair, for as long as I can, so that I can continue to use my body." Sarah expressed a sense of surprise that her accomplishments would not be maintained over the long term. At the relatively young age of 29, she has recognized that she must come to terms with aging with a disability: "I am always on the go, and when I was younger, I could handle that, because aging with a disability was not something I knew much about. So I always presumed that my energy, my mind, and my body would all match. And as a woman, it's not matching anymore."

Self-care was mentioned as being fundamental to these women's sense of well-being. The notion of losing something that was so personally meaningful to the women, and something they had worked so hard to gain through their own perseverance, was difficult. The women acknowledged that there were parts of their lives where support was required (e.g., accessibility of the work place, transportation needs); however, maintaining privacy of person and their home was paramount. To have it removed because of eroding function brought about by pain, loss of strength, or injury created considerable apprehension in the women. Although there was acknowledgement that eventually support with some activities of daily life may be needed in legitimate old age, personal happiness would be sacrificed if it came earlier. Cathy and Leanne express the emotions surrounding the question of personal independence.

> There is never a day when I don't look after my own self-care. That will never happen. Well, I suppose when I'm really old, and then it will be acceptable. I can't imagine not being able to do that. I don't think I would be near as happy with my life if I couldn't look after my self-care. Cathy, age 22

> I'm independent. I need my independence and I think everybody does. I don't think I'm alone in this. We're already dependent enough on people to certain lengths. So to be more dependent on somebody is worse. This might be a strong word, but the best way to describe it would be death. Leanne, age 22

Maintaining Function Through Physical Activity

The women of this study perceived their bodies to be aging sooner than those of women without disabilities. The perception of premature aging was linked, in part, to the presence of pain that was not previously evident. Ironically, the physical challenges the women faced daily were further complicated by pain brought about by either too much or not enough physical activity. Stiffness, joint soreness, and muscle pain was found, in some instances, to be more limiting to their activity levels and overall functioning than their actual disabilities. The women were not only balancing their lifestyle and aging with a disability, they were also managing their pain.

The presence of pain was also a psychological reminder of the downhill changes occurring within their bodies. Feelings of unfairness were expressed as they reflected upon the life issues in their 20s and 30s that women without disabilities have yet to consider. Sarah helps us understand her current struggle with the question of aging and her apprehension for the future.

> I think what I have right now is a gift. But it's not forever. And I am confronting aging earlier and I know why. It's completely related to the disability. And that kind

of scares me, because I don't know what happens when I am legitimately old. I'm talking in a reality that other people my age that don't have disabilities don't talk about. There is pain now, aging with my body and there wasn't before. The pain is what limits me more so than the cerebral palsy. It's the pain, it's aging with the disability . . . and that's disheartening. I mean my whole theme is aging, aging, aging, but for somebody at 29 to be talking like that, what is it saying? Sarah, age 29

Beyond pain, these women foresaw further health challenges resulting from the interception of age and disability. The women wondered how they were going to cope with other illness and their resultant impact on function. Amanda spoke of acquiring secondary disabilities in addition to the balance concerns and muscle weakness that followed her acquired brain injury. She was concerned about the cumulative effect of secondary disabilities on function and expressed concerns about her perceived predisposition to the chronic condition of arthritis.

As you grow older, your body has more disabilities and you have to be more aware of them. If you don't keep up with them, your body will go downhill and you won't be able to move around and do things. I know that by the time I'm 40 or 50 I'm going to have arthritis because I have hurt myself and I've injured certain parts of my body. Amanda, age 37

The reality of chronic diseases that can be argued will affect all people given enough time, are realities in the lives of these young women. The women are aware of the fragility of their long-term health and how decisions made now will impact their futures.

The women who participated in the study considered themselves to be physically active. When asked of the importance of physical activity in their lives, their answers ranged from psychological health, alleviation of pain, and the maintenance of function and independence. Several of the women spoke of the importance of physical activity as an intervention on the impact of aging with a physical disability. Amanda, who has muscular imbalance and weakness, indicated that cardiovascular fitness was less of a priority for her than relaxing and stretching out tight muscles. Amanda explains:

My joints will and have gotten stiffer and stiffer. If I don't exercise them, they won't function. I want to be the best I can be. So I have to maintain my body and my mind. You have to either maintain it or accept that you're going to go downhill. . . Medication isn't always the answer. Yes, it will take away some of the discomfort, but it hides the real problem. You have to go out and do something about it instead of hiding the discomfort with medication. And physically active does not mean you go run ten miles. [For me] it means stretching, just relaxing the body. Amanda, age 37

The women felt that physical activity was important for maintaining hard-earned lifestyles and level of physical functioning. The women were motivated to be physically active because of the dramatic impact of physical inactivity on their function. For Jennifer, who had a spinal cord injury, strength was important.

So physical activity is important in that way to keep my body strong to make up for some problems with the growth in my bones and that kind of thing. Because when I get older, I know even from after my accident my body was going to go downhill faster than the "normal person," you know? So that also kind of makes me want to do more. Jennifer, age 27

For Amanda, physical activity was also an alternate approach to the management of her pain. "Your body falls apart with age and I have to be very aware that because of my accident I have more limitation than other people." Stretching throughout the day to keep her body limber alleviated Amanda's stiffness and pain. She was very motivated to complete her stretches due to the negative consequences otherwise experienced. "I'll go to bed and the next morning I can barely get out of bed because I'm that stiff."

Discussion

The Paradox of Aging with a Disability

The meaning of aging shared by the participants was a complex blend of thoughts, feelings, and actions around the physiological changes experienced and anticipated with age and the increased psychological wisdom associated with accumulated life experiences. The young women experienced an illness reality that represented changes to their body above and beyond those directly attributable to their disabilities and at an age earlier to that of women without disabilities.

The realities of the participants support the work of others who suggested that very subtle age related changes in the systems of the body can have a profound impact on functional ability and independence, or what Vandenakker and Glass (2001) refer to as the disablement process. The participants dispelled the belief that their nonprogressive disabilities were stable after surgery and other medical rehabilitation phases associated with their impairments were completed.

The mean age of the women corroborates the assertion that persons with disabilities may age physiologically sooner than persons without disabilities (Vandenakker & Glass, 2001). The suggestion that unexpected changes in functional ability will occur between 30 and 50 years of age would appear to be accurate given the experiences shared by the participants (Whiteneck et al., 1992). The age of the participants would suggest that the symptoms of pain and fatigue, however, appear even sooner than may be commonly appreciated, potentially precipitating a decrease in physical activity, thereby exacerbating the rate of functional change (Miller et al., 2000). The paradox between the belief that a high quality of life is dependent upon physical wellness and independence and what these young women anticipate for their futures as they age was apparent.

A further paradox may be evident in the social identity of women as they relate to women without disabilities but also maintain a strong disability identity. Sarah's atypical body was perceived to be experiencing a phenomenon shared by all women, bringing about a shared social identity with other women that was self-affirming and normalizing. The perfection of body that was prominent in their self-definition as younger women appeared to dissipate as she integrated the *good* and *bad* pans of their body as women without disabilities around them aged and succumbed to the social values that accompany aging. The existence of this commonality with women without disabilities is in need of further inquiry.

Gill (1997) speaks of the importance of persons with disabilities integrating into the parent culture. The desire to "fit in" and be a member of larger society hinges upon a disability identity that asserts equality because of the unique contribution of a disability perspective, not in spite of it. We heard the women's strong celebration of their lives with disabilities. These women provided a valuable perspective on aging that integrated their life experiences as women and their experiences of impairment and disability. They bridged both worlds in their stories identifying with the value systems of the parent culture (women without disabilities) while also reaffirming their identity with the disability community. Gill (1997) refers to this as coming home or a celebration of life accomplishments and disability identity, at the individual and group level.

The Disablement Process

In their socio-medical model of disablement, Verbrugge and Jette (1994) speak of personal and environmental factors that can speed or slow disablement, namely, risk factors, exacerbators, and interventions.

Fear of loss of function, fear of the unknown, and fear of secondary disabilities were perceived, over time, to be taking away the women's physical freedom. The fear of physical activity lay off due to injury and a permanent change in function as an outcome of injury were a reality for the participants. In its extreme, the fear of injury may inhibit participation in physical activity.

The women of this study recognized the importance of looking after their bodies. Their concerns appeared to revolve around managing the short-term day to day aches, pains, and stiffness that accompanied their daily activities and longer-term concerns about chronic illness. Pain and fatigue were perceived to be antecedents to decreased independence and the need for assistance to complete daily activities. Concerns about chronic illness leading to secondary disabilities, which would rob them of their hard-earned independence, were evident in their reflections.

The potential for decreased independence with age elicited strong psychological feelings about the women's quality of life. The women were disheartened about the prospect of requiring the assistance of others. A depressed mood state has the potential to negatively impact the incentive needed for the women to continue to be causal agents in their own health and physical well-being and maintain physical function as long as possible (Dunn, Trivedi, & O'Neal, 2001; Moore et al., 1999).

The participants clearly indicated the value they placed on physical activity in their lives. It contributed to a sense of physical well-being and was instrumental in their perceptions of health and independence. It was implicated in pain control, the prevention of illness and secondary disability, and the maintenance of function. The women of the study supported the notion that physical activity can play a mediating role in the process of disablement. Although physical decline was anticipated by the women, the rate at which the change occurred and the extent to which it affected their functional lives were felt to be impacted by their commitment to physical activity, whether that be through traditional activities such as swimming and wheeling, or the physical demands that are part of their daily lives.

The under-recognized role of physical activity in the prevention of secondary disabilities in those with early onset disabilities needs to be addressed by health promotion professionals, including adapted physical activity experts. The stories of the women presented clearly highlight the apprehension they feel about their futures and their future quality of life. The immediate health needs of persons with disabilities appear to have taken priority over longer term lifestyle counseling by primary and secondary health care professionals, due in part to our medically oriented and individualistic interpretation of the disablement process (Hughes & Paterson, 1997). The benefits of physical activity on function, pain management, the prevention of secondary chronic illness, and psychological health are important health promotion messages. Further investigations of the experience of aging with a disability and the mediating role of physical activity in the disablement process are encouraged. Results of this study are preliminary and descriptive in nature. Research designs that utilize multiple methods, larger sample sizes, rigorous member checks of findings, and focus in-depth on the phenomenon of aging will add to our current knowledge base.

References

Adams-Price, C.E., Henley, T.B., & Hale, M. (1998). Phenomenology and the meaning of aging for young and old adults. *International Journal of Aging and Human Development*, 47(4), 263-277.

Allen, M.N., & Jensen, L. (1990). Hermeneutic inquiry: Meaning and scope. *Western Journal of Nursing Research*, 12(2), 241-253.

Boult, C., Kane, R.L., Louis, T.A., Boult, L., & McCaffrey, D. (1994). Chronic conditions that lead to functional limitation in the elderly. *Journal of Gerontology: Medical Sciences*, 49, M28-M36.

Carlson, J.E., Ostir, G.V., Black, S.A., Markides, K.S., Rudkin, L., & Goodwin, J.S. (1999). Disability in older adults 2: Physical activity as prevention. *Behavioral Medicine*, 24, 157-168.

Charlifue, S.W., Weitzenkanip, D.A., & Whiteneck, G.G. (1999). Longitudinal outcomes in spinal cord injury: Aging, secondary conditions, and well-being. *Archives of Physical Medicine and Rehabilitation*. 80, 1429-1434.

Cooper, R.A., Quatrano, L.A., Axelson, P.W., Harlan, W., Stinement, M., Franklin, B., et al. (1999). Research on physical activity and health among people with disabilities: A consensus statement. *Journal of Rehabilitation Research and Development*, 36(2), 1-15.

Cresswell, J.W. (1998). *Qualitative inquiry and research design: Choosing among the five traditions.* Thousand Oaks, CA: Sage Publications.

Denzin, N.K. (1994). The art and politics of interpretation. In N.K. Denzin & Y.S. Lincoln (Eds.), *Handbook of qualitative research* (pp. 500-515). Thousand Oaks, CA: Sage Publications.

Dreher, M. (1994). Qualitative research methods from the reviewer's perspective. In J.M. Morse (Ed.), *Critical issues in qualitative research methods* (pp. 281-297). Thousand Oaks. CA: Sage Publications.

Dunlop, D.D., Hughes, S.L., Edelman, P., Singer, R.M., & Chang, R.W. (1998). Impact of joint impairment on disability-specific domains at four years. *Journal of Clinical Epidemiology*, 51, 1253-1261.

Dunn, A.L., Trivedi, M.H., & O'Neal, H.A. (2001). Physical activity dose-response effects on outcomes of depression and anxiety. *Medicine and Science in Sports and Exercise*, 33, S587-597.

Eisenberg, M.G., & Saltz, C.C. (1991). Quality of life among aging spinal cord injured persons: Long term rehabilitation outcomes. *Paraplegia*, 29, 514-520.

Ettinger, W.H. (1998. Physical activity, arthritis, and disability in older people. *Clinics in Geriatric Medicine*, 14, 633-640.

Gerhart, K.A., Bergstrom, E., Charlifue, S.W., Menter, R.R., & Whiteneck, G.G. (1993). Long term spinal cord injury: Functional changes over time. *Archives of Physical Medicine and Rehabilitation*, 74, 1030-1034.

Gill, C. (1997). Four types of integration in disability identity development. *Journal of Vocational Rehabilitation*, 9, 39-46.

Goodwin, D.L., Kuttai, H., & Harris, L. (2001, July). *The experience of the imperfect body: The metaphorical intersection of sickness, disability and physical well-being.* Poster presented at the International Symposium on Adapted Physical Activity, Vienna, Austria.

Hirvensalo, M., Rantanen, T., & Heikkinen, E. (2000). Mobility difficulties and physical activity as predictors of mortality and loss of independence in the community-living older population. *Journal of the American Geriatrics Society, 48,* 493-498.

Hopman-Rock, M., Kraaimaat, F.W., Bijlsam, J.W.J. (1996). Physical activity, physical disability, and osteoarthritic pain in older adults. *Journal of Aging and Physical Activity, 4,* 324-337.

Hootman, J.M., Sniezek, I.E., & Helmick, C.G. (2002). Women and arthritis: Burden, impact, and prevention programs. *Journal of Women's Health and Gender-Based Medicine, 11,* 407-416.

Hughes, B., & Paterson, K. (1997). The social model of disability and the disappearing body: Towards a sociology of impairment. *Disability and Society, 12,* 325-340.

Jackson, J.E. (1990). "I am a field note:" Fieldnotes as a symbol of professional identity. In R. Sanjek (Ed.), *Fieldnotes: The makings of anthropology* (pp. 3-33). Cornell, NY: Cornell University Press.

Kavanagh, T., & Shephard, R.J. (1990). Can regular sports participation slow the aging process? Data on masters athletes. *The Physician and Sportsmedicine, 18(0),* 94-103.

Keller, M.L., Leventhal, E.A., & Larson, B. (1989). Aging: The lived experience. *International Journal of Aging and Human Development, 29,* 67-82.

Kemp, B., & Miosqueda, L. (1997). Aging-related conditions. In M. Fuhrer (Ed.). *Assessing medical rehabilitation practices: The promise of outcomes research* (pp. 393-411). New York: Brooks Publishing.

Kvale, S. (1996). *Interviews: An introduction to qualitative research interviewing.* Thousand Oaks, CA: Sage Publications.

Lal, S. (1998). Premature degenerative shoulder changes in spinal cord injury patients. *Spinal Cord, 36,* 186-189.

Lammertse, D.P., & Yarkony, G.M. (1991). Rehabilitation in spinal cord disorders: Outcomes and issues of aging after spinal cord injury. *Archives of Physical Medicine and Rehabilitation, 72,* S309-S311.

Lawrence, R.H., & Jette, A.M. (1996). Disentangling the disablement process. *Journal of Gerontology, 51B(4),* S173-S182.

Lincoln, Y.S., & Guba, E.G. (1985). *Naturalistic inquiry.* Newbury Park, CA: Sage Publications.

Meadows, L.M., & Morse, J.M. (2001). Constructing evidence within the qualitative project. In J.M. Morse, J.M. Swanson, & A.J. Kuzel (Eds.), *The nature of qualitative evidence* (pp. 187-202). Thousand Oaks, CA: Sage Publications.

Miller, ME., Rejeski, W.J., Reboussin, B.A., Ten-Have, T.R., & Ettinger, W.H. (2000). Physical activity, functional limitations, and disability in older adults. *Journal of the American Geriatrics Society, 48,* 1264-1272,

Montepare, J.M., & Lachman, M.E. (1989). "You're only as old as you feel:" Self-perceptions of age, fears of aging, and life satisfaction from adolescence to old age. *Psychology of Aging, 4(1),* 73-78.

Moore, K.A., Babyak, M.A., Wood, C.E., Napolitano, M.A., Kharti, P., Craighead, W.E. et al. (1999). The association between physical activity and depression in older depressed adults. *Journal of Aging and Physical Activity, 71,* 55-61.

Moustakas, C. (1994). *Phenomenological research methods.* Thousand Oaks, CA: Sage Publications.

Nagi. S.Z. (1965). Some conceptual issues in disability and rehabilitation. In M.B. Sussman (Ed.), *Sociology and rehabilitation* (pp. 100-113). Washington, DC: American Sociological Association.

Paterson, K., & Hughes, B. (1999). Disability studies and phenomenology: The carnal politics of everyday life. *Disability and Society, 14,* 598-610.

Patton, M.Q. (1997). *Qualitative evaluation and research methods* (3rd ed.). Thousand Oaks, CA: Sage Publications.

Pentland, W., McColl, M.A., & Rosenthal, C. (1995). The effect of aging and duration of disability on long term health outcomes following spinal cord injury. *Paraplegia, 33,* 367-373.

Peters, D.J. (1995). Human experience in disablement: The imperative of the ICIDH. *Disability and Rehabilitation,* 17(3/4), 135-144.

Peters, D.J. (1996). Disablement observed, addressed, and experienced: Integrating subjective experience into disablement models. *Disability and Rehabilitation, 18,* 593-603.

Rantanen, T., Guralnik, J.M., Sakari-Rantala, R., Leveille. S., Simonsick, E.M., Ling, S. et al. (1999). Disability, physical activity, and muscle strength in older women: The women's health and aging study. *Archives of Physical Medicine and Rehabilitation, 80,* 130-135.

Rejeski, W.J., & Focht, B.C. (2002). Aging and physical disability: On integrating group and individual counseling with the promotion of physical activity. *Exercise and Sports Sciences Reviews,* 30(4), 166-170.

Schofield, J.W. (1990). Increasing the generalizability of qualitative research. In E.W. Eisner & A. Peshkin (Eds.), *Qualitative inquiry in education* (pp. 171-233). New York: Columbia University Press.

van Manen, M. (1997). *Researching lived experience: Human science for an action sensitive pedagogy.* London, ON: The Althouse Press.

Vandenakker, C.B., & Glass, D.D. (2001). Menopause and aging with disability. *Physical Medicine and Rehabilitation Clinics of North America,* 12(1), 133-151.

Verbrugge, L.M., & Jette, A.M. (1994). The disablement process. *Social Science and Medicine,* 38(1), 1-14.

Wheeler, G.D., Malone, L.A., VanVlack, S., Nelson, E., & Steadward, R.D. (1996). Retirement from disability sport: A pilot study. *Adapted Physical Activity Quarterly, 13,* 382-399.

Whiteneck, G.G., Charlifue, S.W., Frankel. H.L., Fraser, M.H., Garder, B.P., Gerhart, K.A., et al. (1992). Mortality, morbidity, and psychosocial outcomes of persons spinal cord injured more than 20 years ago. *Paraplegia, 30,* 617-630.

Wolcott, H. (1994). *Transforming qualitative data: Description, analysis, and interpretation.* Thousand Oaks, CA: Sage Publications.

Wolcott, H.F. (2001). *Writing up qualitative research* (2nd ed.). Thousand Oaks, CA: Sage Publications.

Author Note

This study was funded by the Health Science Utilization Research Commission of Saskatchewan, Canada.

Appendix D

Answers for "Checking Your Knowledge" Sections

Chapter 1	Chapter 2	Chapter 3	Chapter 4	Chapter 5
1. b	1. c	1. e	1. b	1. a
2. c	2. b	2. b	2. e	2. b
3. b	3. b	3. a	3. a	3. c
4. a	4. a	4. c	4. a	4. a
5. e		5. b	5. c	5. c
6. d			6. d	

Chapter 6	Chapter 7	Chapter 8
1. b	1. c	1. b
2. a	2. b	2. c
3. d	3. b	3. d
4. e	4. e	4. b
5. d	5. d	5. d

Chapter 9	Chapter 10	Chapter 11
1. a	1. a	1. a
2. c	2. d	2. b
3. a	3. b	3. b
4. b	4. b	4. b
5. b	5. a	5. a
6. a	6. d	
7. d		
8. c		

References

References for material cited in appendix A can be found on pages 166-168; in appendix B, pages 183-185; in appendix C, pages 200-202.

Aita, M., & Richer, M. (2005). Essentials of research ethics for healthcare professionals. *Nursing and Health Sciences, 7,* 119–125.

Alexa, M., & Zuell, C. (2000). Text Analysis Software: Commonalities, Differences and limitations: The Results of a Review. *Quality and Quantity, 34*(3), 299-321.

Allen, J., & Howe, B. (1998). Player ability, coach feedback, and female adolescent athletes' perceived competence and satisfaction. *Journal of Sport and Exercise Psychology, 20,* 280-299.

American Medical Association. (1998). *Manual of style: A guide for authors and editors* (9th ed.). Philadelphia: Lippincott Williams & Williams.

American Psychological Association. (2001). *Publication manual of the American Psychological Association* (5th ed.). Washington, DC: American Psychological Association.

Amorose, A., & Horn, T. (2000). Intrinsic motivation: Relationships with collegiate athletes' gender, scholarship status, and perceptions of their coaches' behavior. *Journal of Sport and Exercise Psychology, 22,* 63-84.

Annells, M. (1996). Grounded theory method: Philosophical perspectives, paradigm of inquiry, and postmodernism. *Qualitative Health Research, 6,* 379-394.

Ary, D., Cheser Jacobs, L.C., & Razavieh, A. (2002). *Introduction to research in education* (6th ed.). Belmont, CA: Wadsworth.

Bain, L.L. (1989). Interpretive and critical research in sport and physical education. *Research Quarterly for Exercise and Sport, 60*(1), 21-24.

Ball, K., Salmon, J., Giles-Corti, B., & Crawford, D. (2006). How can socio-economic differences in physical activity among women be explained? A qualitative study. *Women & Health, 43*(1), 93-113.

Bandura, A. (1977). Self-efficacy: Toward a unifying theory of behavior change. *Psychological Review, 84,* 191- 215.

Barbour, R.S. (2001). Checklists for improving rigour in qualitative research: A case of the tail wagging the dog? *British Medical Journal, 322,* 1115-1117.

Barry, C. (1998). Choosing Qualitative Data Analysis Software: Atlas/ti and Nudist Compared. *Sociological Research Online, 3*(3), http://www.socresonline.org.uk/socresonline/3/3/4.html.

Baszanger, I., & Dodier, N. (1997). Ethnography: Relating the part to the whole. In D. Silverman (Ed.), *Qualitative research: Theory, method and practice.* Thousand Oaks, CA: Sage.

Beach, D. (2005). From fieldwork to theory and representation in ethnography. In G. Troman, B. Jeffrey, & G. Walford (Eds.), *Methodological issues and practices in ethnography* (pp. 1-17). Boston: Elsevier.

Berger, P.L., & Luckmann, T. (1966). *The social construction of reality: A treatise in the sociology of knowledge.* Garden City, NY: Doubleday.

Black, S., & Weiss, M. (1992). The relationship among perceived coaching behaviors, perceptions of ability, and motivation in competitive age-group swimmers. *Journal of Sport and Exercise Psychology, 14*, 309-325.

Bluff, R. (2005). Grounded theory: The methodology. In I. Holloway (Ed.), *Qualitative research in health care* (pp. 147-165). Berkshire, England: Open University Press.

Bogdan, R.C., & Biklen, S.K. (2003). *Qualitative research for education: An introduction to theories and methods* (4th ed.). Boston: Allyn and Bacon.

Bogdan, R.C., & Biklen, S.K. (2007). *Qualitative research for education: An introduction to theories and methods* (5th ed.). Boston: Allyn and Bacon.

Booth, W.C., Colomb, G.C., & Williams, J.M. (2003). *The craft of research* (2nd ed.). Chicago: University of Chicago.

Borseth, K.M. (2004). *An investigation of the social support network of injured athletes.* Unpublished master's thesis, Northern Illinois University, De Kalb, Illinois.

Byrne, M.M. (2001). Understanding life experiences through a phenomenological approach to research. *AORN Journal, 73*, 830-832.

Chenitz, W.C., & Swanson, J.M. (1986). Qualitative research using grounded theory. In W.C. Chenitz & J.M. Swanson (Eds.). *From practice to grounded theory* (pp. 3-15). Menlo Park, CA: Addison Wesley.

Chiang, L. (2005). Exploring the health-related quality of life among children with moderate asthma. *Journal of Nursing Research, 13*(1), 31-39.

Colaizzi, P.F. (1978). Psychological research as the phenomenologist views it. In R.S. Valle & M. King (Eds.). *Existential-phenomenological alternatives for psychology.* New York: Oxford University Press.

Cortozzi, M. (1993). *Narrative analysis.* Bristol, PA: Falmer Press.

Coyne, I.T. (1997). Sampling in qualitative research. Purposeful and theoretical sampling; merging or clear boundaries. *Journal of Advanced Nursing, 26*, 623-630.

Creswell, J.W. (2007). *Qualitative inquiry & research design: Choosing among five approaches* (2nd ed.). Thousand Oaks, CA: Sage.

Creswell, J.W. (2005). *Educational research: Planning, conducting, and evaluating quantitative and qualitative research* (2nd ed.). Upper Saddle River, NJ: Pearson Merrill Prentice Hall.

Creswell, J.W. (2003). *Research design: Qualitative, quantitative, and mixed methods approaches* (2nd ed.). Thousand Oaks, CA: Sage.

Creswell, J.W. (1998). *Qualitative inquiry and research design: Choosing among five traditions.* Thousand Oaks, CA: Sage.

Denscombe, M. (1998). *The good research guide for small scale research projects.* Buckingham, England: Open University Press.

Denzin, N.K., & Lincoln, Y.S. (Eds.). (2000). *Handbook of qualitative research* (2nd ed.). Thousand Oaks, CA: Sage.

Devers, K.J., & Frankel, R.M. (2000). Study design in qualitative research-2: Sampling and data collection strategies. *Education for Health, 13*(2), 263-271.

Dingwall, R. (1992). Don't mind him—he's from Barcelona: Qualitative methods in health studies. In J. Daly, I. MacDonald, & E. Willis (Eds.). *Researching health care: Design, dilemmas, disciplines* (pp. 161-175). London: Tavistoch/Routledge.

Dodd, K.J., Taylor, N.F., Denisenko, S., & Prasad, D. (2006). A qualitative analysis of a progressive resistance exercise programme for people with multiple sclerosis. *Disability and Rehabilitation, 28*(18),1127-1134.

Dudgeon, B.J., Gerrard, B.C., Jensen, M.P., Rhodes, L.A., & Tyler, E.J. (2002). Physical disability and the experience of chronic pain. *Archives of Physical Medicine and Rehabilitation, 83*, 229-235.

Dunn, A.L., Trivedi, M.H., & O'Neil, H.A. (2001). Physical activity dose-response effects on outcomes of depression and anxiety. *Medicine and Science in Sports and Exercise, 33,* S587-S597.

Erlandson, D.A., Harris, E.L., Skipper, B.L., & Allan, S.D. (1993). *Doing naturalistic inquiry: A guide to methods.* Thousand Oaks, CA: Sage.

Evans, J., & Penney, D. (2008). Levels on the playing field: The social construction of physical 'ability' in the physical education curriculum. *Physical Education & Sport Pedagogy, 13*(1), 31-47.

Fowkes, F.G.R., & Fulton, P.M. (1991). Critical appraisal of published research: Introductory guidelines. *British Medical Journal, 302,* 1136-1140.

Gilbert, M.J., & Fleming, M.F. (2006). Pediatricians' approach to obesity prevention counseling with their patients. *Wisconsin Medical Journal, 105*(5), 26-31.

Giorgi, A. (1985). *Phenomenology and psychological research.* Pittsburgh: Duquesne University Press.

Glaser, B. (1978). *Theoretical sensitivity: Advances in the methodology of grounded theory.* Mill Valley, CA: Sociology Press.

Glaser, B.G., & Strauss, A.L. (1967). *The discovery of grounded theory.* Hawthorne, NY: Aldine.

Golden-Biddle, K., & Locke, K. (2007). *Composing qualitative research* (2nd ed.). Thousand Oaks, CA: Sage.

Goodwin, D.L., & Compton, S.G. (2004). Physical activity experiences of women aging with disabilities. *Adapted Physical Activity Quarterly, 21,* 122-138.

Goodwin, D.L., Thurmeier, R., & Gustafson, P. (2004). Reactions to the metaphors of disability: The mediating effects of physical activity. *Adapted Physical Activity Quarterly, 21,* 379-398.

Graham, R.C., Dugdill, L., & Cable, N.T. (2005). Health professionals' perspectives in exercise referral: Implications for the referral process. *Ergonomics, 48,* 1411-1422.

Guba, E.G. (1981). Criteria for assessing the trustworthiness of naturalistic inquiries. *Educational Communication and Technology Journal, 29,* 75-92.

Hallberg, L.R.M. (2006). The "core category" of grounded theory: Making constant comparisons. *International Journal of Qualitative Studies on Health and Well-being, 1,* 141-148.

Harris, J.C. (1983). Broadening horizons: Interpretive cultural research, hermeneutics, and scholarly inquiry in physical education. *Quest, 35*(2), 82-95.

Hayes, N. (1997). Introduction: Qualitative research and research in psychology. In N. Hayes (Ed.), *Doing qualitative analysis in psychology* (pp. 1-9). Hove, England: Psychology Press.

Hesketh, K., Waters, E., Green, J., Salmon, L., & Williams, J. (2005). Healthy eating, activity and obesity prevention: A qualitative study of parent and child perceptions in Australia. *Health Promotion International, 20*(1), 19-26.

Higginbottom, G.M.A. (2004). Sampling issues in qualitative research. *Nurse Researcher, 12*(1), 7-19.

Hodge, S.R., Tannehill, D., & Kluge, M.A. (2003). Exploring the meaning of practicum experiences for PETE students. *Adapted Physical Activity Quarterly, 20,* 381-399.

Holloway, I. (Ed.). (2005). *Qualitative research in health care.* Berkshire, England: Open University Press.

Horn, T.S. (1984). Expectancy effects in the interscholastic athletic setting. *Journal of Sport Psychology, 6,* 60-76.

Horsburgh, D. (2003). Evaluation of qualitative research. *Journal of Clinical Nursing, 12,* 307-312.

Hunter, C.L., Spence, K., McKenna, K., & Iedema, R. (2008). Learning how we learn: An ethnographic study in a neonatal intensive care unit. *Journal of Advanced Nursing, 62*(6), 657–664.

Jeffers, B.R. (2002). Continuing education in research ethics for the clinical nurse. *Journal of Continuing Education in Nursing, 33*(6), 265-269.

Jette, D., Bertoni, A., Coots, R., Johnson, H., McLaughlin, C., & Weisbach, C. (2007). Clinical instructors' perceptions of behaviors that comprise entry-level clinical performance in physical therapist students: A qualitative study. *Physical Therapy, 87*(7), 833-843.

Kendall, J. (1999). Axial coding and the grounded theory controversy. *Western Journal of Nursing Research, 21*, 743-757.

Kenow, L., & Williams, J. (1999). Coach-athlete compatibility and athletes' perceptions of coaching behaviors. *Journal of Sport Behavior, 22*, 251-259.

Khunti, K., Stone, M.A., Bankart, J., Sinfield, P., Pancholi, A., Walker, S., et al. (2007). Primary prevention of type-2 diabetes and heart disease: Action research in secondary schools serving an ethnically diverse UK population. *Journal of Public Health, 30*(1), 30-37.

Knight, K.L., & Ingersoll, C.D. (1996). Optimizing scholarly communications: 30 tips for writing clearly. *Journal of Athletic Training, 31*, 209-213.

Knight, K.L., & Ingersoll, C.D. (1996). Structure of a scholarly manuscript: 66 tips for what goes where. *Journal of Athletic Training, 31*, 201-206.

Kvale, S. (1996). *InterViews: An introduction to qualitative research interviewing.* Thousand Oaks, CA: Sage.

Lancy, D.F. (1993). *Qualitative research in education: An introduction to the major traditions.* White Plains, NY: Longman.

Lazarton, A. (2003). Evaluative criteria for qualitative research in applied linguistics: Whose criteria and whose research? *The Modern Language Journal, 87*, 1-12.

Lemon, N., & Taylor, H. (1997). Caring in casualty: The phenomenology of nursing care. In N. Hayes (Ed.), *Doing Qualitative Analysis in Psychology* (pp. 227-244). Hove, England: Psychology Press.

Lempp, H., & Seale, C. (2004). The hidden curriculum in undergraduate medical education: Qualitative study of medical students' perceptions of teaching. *British Medical Journal, 329*, 770-773.

Lincoln, Y.S. (1995). Emerging criteria for quality in qualitative and interpretive research. *Qualitative Inquiry, 1*, 275-289.

Lincoln, Y.S., & Guba, E.G. (1985). *Naturalistic inquiry.* Thousand Oaks, CA: Sage.

Locke, L.F. (1989). Qualitative research as a form of scientific inquiry in sport and physical education. *Research Quarterly for Exercise and Sport, 60*(1), 1-20.

Locke, L.F., Silverman, S.J., & Spirduso, W.W. (1998). *Reading and understanding research.* Thousand Oaks, CA: Sage.

Locke, L.F., Spirduso, W.W., & Silverman, S.J. (2000). *Proposals that work: A guide for planning dissertation and grant proposals* (4th ed.). Thousand Oaks, CA: Sage.

Loeb, S., Penrod, J., & Hupcey, J. (2006). Focus groups and older adults: Tactics for success. *Journal of Gerontological Nursing, 32*(3), 32-38.

McMillan, J. & Wergin, J. (2005). Understanding and evaluating educational research (2nd Ed.). Upper Saddle River, NJ: Merrill.

Malasarn, R., Bloom, G.A., & Crumpton, R. (2002). The development of expert male National Collegiate Athletic Association Division I certified athletic trainers. *Journal of Athletic Training, 37*, 55-62.

Malterud, K. (2001). Qualitative research: Standards, challenges, and guidelines. *The Lancet, 358*, 483-488.

Marshall,C., & Rossman, G.B. (1999). *Designing qualitative research* (3rd ed.). Thousand Oaks, CA: Sage.

Martens, R. (1987). Science, knowledge, and sport psychology. *The Sport Psychologist, 1*(1), 29-55.

Maxwell, J.A. (1996). *Qualitative research design: An interactive approach.* Thousand Oaks, CA: Sage.

Mays, N., & Pope, C. (1995). Qualitative research: Rigour and qualitative research. *British Medical Journal, 311*, 109-112.

McGarvey, H.E., Chambers, M.G., & Boore, J.R.P. (2004). The influence of context on role behaviors of perioperative nurses. *Associate of Operating Room Nurses Journal, 80*(6), 1103-1119.

Mensch, J.M., & Ennis, C.D. (2002). Pedagogic strategies perceived to enhance student learning in athletic training education. *Journal of Athletic Training, 37,* S199-S207.

Merriam, S.B. (1998). *Qualitative research and case study applications in education* (2nd ed.). San Francisco: Jossey-Bass.

Merriam, S.B. (Ed.). (2002). *Qualitative research in practice: Examples for discussion and analysis.* San Francisco: Jossey-Bass.

Miles, M.B., & Huberman, M. (1984). *Qualitative data analysis: A sourcebook of new methods.* Thousand Oaks, CA: Sage.

Mills, G.E. (2007). *Action research: A guide for the teacher researcher* (3rd ed.). Upper Saddle River, NJ: Pearson Merrill Prentice Hall.

Mills, G.E. (2003). *Action research: A guide for the teacher researcher* (2nd ed.). Upper Saddle River, NJ: Merrill Prentice Hall.

Moore, K.A., Babyak, M.A., Wood, C.E., Napolitano, M.A., Kharti, P., Craighead, W.E., et al. (1999). The association between physical activity and depression in older depressed adults. *Journal of Aging and Physical Activity, 71,* 55-61.

Morison, M., & Moir, J. (1998). The role of computer software in the analysis of qualitative data: Efficient clerk, research assistant, or Trojan horse? *Journal of Advanced Nursing, 28*(1), 106-116.

Morse, J.M. (1991). *Qualitative nursing research: A contemporary dialogue.* Newbury Park, CA: Sage.

Morse, J.M. (2000). Determining sample size. *Qualitative Health Research, 10,* 3-5.

Moustakas, C. (1994). *Phenomenological research methods.* Thousand Oaks, CA: Sage.

Munhall, P.L., & Boyd, C.O. (1993). *Nursing research: A qualitative perspective* (2nd ed.). New York: National League for Nursing.

Noble-Adams, R. (1999). Ethics and nursing research 1: Development, theories, and principles. *British Journal of Nursing, 8*(13), 888-892.

Oldfather, P., & West, J. (1994). Qualitative research as jazz. *Educational Researcher, 23*(8), 22-26.

Pandit, N.R. (1996). The creation of grounded theory: A recent application of the grounded theory method. *The Qualitative Report, 2*(4). Retrieved November 11, 2008 from http://www.nova.edu/ssss/QR/QR2-2/pandit.html

Parse, R.R. (2001). *Qualitative inquiry: The path of sciencing.* Sudbury, MA: Jones and Bartlett.

Paul, J.L. (2005). *Introduction to the philosophies of research and criticism in education and the social sciences.* Upper Saddle River, NJ: Pearson Merrill Prentice Hall.

Patten, M.L. (1997). *Understanding research methods: An overview of the essentials.* Los Angeles: Pyrczak.

Patton, M.Q. (2002). *Qualitative research & evaluation methods* (3rd ed.). Thousand Oaks, CA: Sage.

Patton, M.Q. (1999). Enhancing the quality and credibility of qualitative analysis. *Health Services Research, 34,* 1189-1208.

Patton, M.Q. (1990). *Practical evaluation.* Newbury Park, CA: Sage.

Pitney, W.A. (2008). *Maintaining professional commitment among athletic trainers working in the high school setting.* A paper presented at the Great Lakes Athletic Trainers' Association Winter Meeting & Clinical Symposium. Toledo, OH.

Pitney, W.A. (2006). Organizational influences and quality-of-life issues during the professional socialization of certified athletic trainers working in the National Collegiate Athletic Association Division I setting. *Journal of Athletic Training, 41*(2), 189-194.

Pitney, W.A. (2002). The professional socialization of certified athletic trainers in high school settings: A grounded theory investigation. *Journal of Athletic Training, 37*(3), 286-292.

Pitney, W.A., & Parker, J. (2001). Qualitative inquiry in athletic training: Principles, possibilities, and promises. *Journal of Athletic Training, 36*(2), 185-189.

Pitney, W.A., & Parker, J. (2002). Qualitative research applications in athletic training. *Journal of Athletic Training, 37*(4 Supplement), S-168-S-173.

Pitney, W.A. (2004). Strategies for establishing trustworthiness in qualitative research. *Athletic Therapy Today, 9*(1), 45-47.

Pitney, W.A., & Ehlers, G.G. (2004). A grounded theory study of the mentoring process involved with undergraduate athletic training students. *Journal of Athletic Training, 39*(4), 344-351.

Pitney, W.A., Stuart, M.E., & Parker, J. (2008). Role strain among dual position physical educators and athletic trainers working in the high school setting. *The Physical Educator, 65*(3), 157-168.

Pizzari, T., McBurney, H., Taylor, N.F., & Feller, J.A. (2002). Adherence to anterior cruciate ligament rehabilitation: A qualitative analysis. *Journal of Sport Rehabilitation, 11*, 90-102.

Podlog, L., & Eklund, R.C. (2006). A longitudinal investigation of competitive athletes' return to sport following serious injury. *Journal of Applied Sport Psychology, 18*, 44-68.

Pope, C.C., & O'Sullivan, M. (2003). Darwinism in the gym. *Journal of Teaching in Physical Education, 22*, 311-327.

Quinn, S.C. (2004). Ethics in public health research: Protecting human subjects: The role of community advisory boards. *American Journal of Public Health, 94*(6), 918-922.

Reed, S., & Giacobbi, P.R. (2004). The stress and coping response of certified graduate athletic training students. *Journal of Athletic Training, 39*, 193-200.

Richardson, L., & St. Pierre, E.A. (2005). Writing: A method of inquiry. In N.K. Denzin & Y.S. Lincoln (Eds.), *The Sage handbook of qualitative research* (3rd ed., pp. 959–978). Thousand Oaks, CA: Sage.

Sage, G.H. (1989). A commentary on qualitative research as a form of scientific inquiry in sport and physical education. *Research Quarterly for Exercise and Sport, 60*(1), 25-29.

Schatzman, L., & Strauss, A.L. (1973). *Field research: Strategies for a natural sociology.* Englewood Cliffs, NJ: Prentice-Hall.

Schmid, T. (2004). Meanings of creativity within occupational therapy practice. *Australian Occupational Therapy Journal, 51*, 80-88.

Schostak, J. (2006). *Interviewing and representation in qualitative research.*

Schram, T.H. (2006). *Conceptualizing and proposing qualitative research* (2nd ed.). Upper Saddle River, NJ: Pearson Merrill Prentice Hall.

Schutz, R.W. (1989). Qualitative research: Comments and controversies. *Research Quarterly for Exercise and Sport, 60*(1), 30-35.

Schwandt, T.A. (1997). *Qualitative inquiry: A dictionary of terms.* Thousand Oaks, CA: Sage.

Seidman, I.E. (1991). *Interviewing as qualitative research: A guide for researchers in education and the social sciences.* New York: Teachers College Press.

Shank, G.D. (2006). *Qualitative research: A personal skills approach* (2nd ed.). Upper Saddle River, NJ: Pearson Merrill Prentice Hall.

Sharkey, S., & Larsen, J.A. (2005). Ethnographic exploration: Participation and meaning in everyday life. In I. Holloway (Ed.), *Qualitative Research in Health Care* (pp. 168-190). Berkshire, England: Open University Press.

Siedentop, D. (1989). Do the lockers really smell? *Research Quarterly for Exercise and Sport, 60*(1), 36-41.

Silverman, D. (2000). *Doing qualitative research: A practical handbook.* Thousand Oaks, CA: Sage.

Spradley, J.P. (1980). *Participant observation.* New York: Holt, Rinehart and Winston.

Sparkes, A.C. (2001). Myth 94: Qualitative health researchers will agree about validity. *Qualitative Health Research, 11*(4), 538-552.

Stanley, B.H., Sieber, J.E., & Melton, G.B. (1996). *Research ethics: A psychological approach.* Lincoln, NE: University of Nebraska Press.

Strauss, A.L., & Corbin, J.M. (1990). *Basics of qualitative research: Grounded theory procedures and techniques.* Newbury Park, CA: Sage.

Stringer, E. (2004). *Action research in education.* Upper Saddle River, NJ: Pearson Merrill Prentice Hall.

Thompson, A., Humbert, M., & Mirwald, R. (2003). A longitudinal study of the impact of childhood and adolescent physical activity experiences on adult physical activity perceptions and behaviors. *Qualitative Health Research, 13*(3), 358-377.

Tjeerdsma, B. L. (1995). If-then statements help novice teachers deal with the unexpected. *Journal of Physical Education, Recreation and Dance, 66*(9), 22-24.

Todres, L. (2005). Clarifying the life-world: Descriptive phenomenology. In I. Holloway (Ed.), *Qualitative Research in Health Care* (pp. 104-124). Berkshire, England: Open University Press.

Tuckett, A. (2004). Qualitative research sampling: The very real complexities. *Nurse Researcher, 12*(1), 47-61.

Udermann, B.E., Schutte, G.E., Reineke, D.M., Pitney, W.A., & Gibson, M.H. (2008). An asessment of spirituality in the curricula of accredited athletic training education programs. *Athletic Training Education Journal, 3*(1), 21-27.

Walker, S.E., Pitney, W.A., Lauber, C., & Berry, D.C. (2005, March). *The perceptions of certified athletic trainers toward continuing education.* Paper presented at the Great Lakes Athletic Trainers' Association Winter Meeting. Toledo, OH.

Webb, C. (1999). Analysing qualitative data: Computerized and other approaches. *Journal of Advanced Nursing, 29,* 323-330.

Whittemore, R., Chase, S.K., & Mandle, C.L. (2001). Validity in qualitative research. *Qualitative Health Research, 11*(4), 522-537.

Wimpeny, P., & Gass, J. (2000). Interviewing in phenomenology and grounded theory: Is there a difference? *Journal of Advanced Nursing, 31*(6), 1485-1492.

Young, K., White, P., & McTeer, W. (1994). Body talk: Male athletes reflect on sport, injury, and pain. *Sport Sociology Journal, 11*(2), 175-194.

Index

Note: The italicized *f* and *t* following page numbers refer to figures and tables, respectively.

About the Authors

William A. Pitney, EdD, ATC, is an associate professor in the department of kinesiology and physical education at Northern Illinois University. Dr. Pitney is a recognized leader in qualitative research in the athletic training profession. He has authored more than 20 peer-reviewed articles and two textbooks and is a section editor for the *Journal of Athletic Training*, in which he published one of the first articles on qualitative research. He is also the associate editor for the *Athletic Training Education Journal* and has served on the Great Lakes Athletic Trainers' Association Research Assistance Committee.

Dr. Pitney earned a bachelor's degree in physical education with a specialization in athletic training from Indiana State University in 1988, a master's degree in physical education from Eastern Michigan University in 1992, and an EdD in adult continuing education from Northern Illinois University in 2000. In his leisure time, he enjoys mountaineering, bicycling, and running.

Jenny Parker, EdD, is an associate professor in the department of kinesiology and physical education at Northern Illinois University. Dr. Parker has published numerous peer-reviewed articles and two book chapters and has presented at the state, regional, and national levels. She is a reviewer for *Research Quarterly for Exercise and Sport*, and she has obtained both internal and external research and instructional grant funding. Her experience with qualitative, quantitative, and mixed-method research studies greatly aided in shaping the book, as did her teaching expertise in providing pedagogical aspects and learning activities.

Dr. Parker earned her bachelor's degree in physical education at the College of St. Paul and St. Mary in England and earned her master's degree in physical education teaching analysis from the University of Oregon in 1991. She earned her EdD in physical education teacher education from the University of Massachusetts in 1996. She received an Outstanding Educator Award in Education from Northern Illinois University in 2008 and has been recognized nationally for her mentoring of undergraduate and graduate physical education students.